Introduction to Groups and Geometries

Introduction to Groups and Geometries

David W. Lyons

Lebanon Valley College
Annville, PA, USA

February 2026 Edition

About the author David W. Lyons is a professor of mathematics at Lebanon Valley College in Annville, Pennsylvania, USA, where he has led a student-faculty research program in quantum information science since 2002.

Visit Lyons' academic website at `quantum.lvc.edu/lyons`.

Cover design: Jean Tashima
ISBN 978-1-958469-36-1 (hardcover)
ISBN 978-1-958469-37-8 (softcover)
Published by 619Wreath

Preface

The purpose of this text is to take advantage of the overlap between introductory courses in group theory and modern geometry. Group theory typically features geometric content in the form of symmetry groups, and Kleinian geometry relies on the group structure of congruence transformations. Learning the two subjects together enhances both.

This text is two textbooks in one: an introduction to group theory, and an introduction to modern geometries using the Kleinian paradigm. The book can be used for a combined one-semester course in both subjects, or, through supplementary projects, it can be used for a one-semester introduction to group theory or a one-semester introduction to modern geometries.

The chapter on groups develops the basic vocabulary and theory of groups and homomorphisms, culminating with group actions. The chapter on geometry makes use of group symmetries to build the basic theory of Möbius, hyperbolic, elliptic, and projective geometries. Throughout, a design theme is the use of a small but carefully chosen collection of tools, beginning with the algebra and geometry of complex numbers and quaternions, and using a minimum of machinery from analysis and linear algebra, to develop useful and nontrivial results for group theory and geometry in a way that prefers using conceptual tools over brute computation.

Many results (for example, Lagrange's Theorem for groups and area formulas in hyperbolic and elliptic geometries) are developed in carefully structured exercises, rather than in the reading. This reflects a deliberate emphasis on active engagement with the material. The intention is for students to read, to reason, and to develop results on their own, as a means of achieving proficiency and fostering analytic skills.

Note on reading exercises versus end-of-section exercises: In the narrative for each section, exercises labeled "Checkpoint" are meant to be reading exercises, that is, part of the process of active reading. Working through end-of-section exercises is meant to take place after reading and working the reading exercises.

The text assumes prerequisite courses in calculus, linear algebra, and experience with proof writing.

Here are sample schedules for three possible courses in a 15-week semester. Variations are supported by "Additional Exercises" sections in Chapters 2 and 3, and a "Further Topics" section at the end of the text.

Table 0.0.1 Combined Course on Groups and Geometries

Ch. 1 Preliminaries	1.1 — 1.4	3 weeks
Ch. 2 Groups	2.1 — 2.5	6 weeks
Ch. 3 Geometries	3.1 — 3.5	6 weeks

Table 0.0.2 Stand-alone Course on Group Theory

Ch. 1 Preliminaries	1.1,1.4	1.5 weeks
Ch. 2 Groups	2.1 — 2.6	12 weeks
Final Project		1.5 weeks

Table 0.0.3 Stand-alone Course on Modern Geometries

Ch. 1 Preliminaries	1.2,1.3	1.5 weeks
Ch. 3 Geometries	3.1 — 3.6	12 weeks
Final Project		1.5 weeks

Many thanks to my readers and problem checkers!

Joshua Miller, Dakota Johnson-Ortiz, Alex Heilman, Ashley Swogger, Jesse Arnold, Daniel Mannetta, Travis Martin, Tyler Hoover, Qinhao Jin, Tyler Pick, Adam Rilatt, Luke Bakalyar, Brandon Bauer, Jack Putnam, Penn Smith, Richard Hammack, Turner Hannon, Alex Baver, Hayden Daubert, Olivia Delgiacco, Justin DeShong, Emily Miller, Hannah Moran

Comments, discussion, error reports, and so on, are welcome.

Note on colors in figures: Some figures in this text use colors. These are viewable in the online version of this text at `mathvista.org`.

Contents

Appendices

Back Matter

Chapter 1

Preliminaries

1.1 The Complex Plane

The complex numbers were originally invented to solve problems in algebra. It was later recognized that the algebra of complex numbers provides an elegant set of tools for geometry in the plane. For an introduction (or for a review) of the basics of the algebra and geometry of the complex numbers, we refer the reader to the section entitled Complex Numbers (`mathvista.org/not_just_calculus/complex_plane_section.html`) in the author's text *Not Just Calculus* [5]. The remainder of this short section introduces material that will be useful later in this text.

Circles and Lines. Let C be the circle in the complex plane with radius $r > 0$ and with center $a \in \mathbb{C}$. A point z lies on C if and only if the distance from z to a equals r. In mathematical symbols, C is the set of complex solutions z for the following equation.

$$|z - a| = r \tag{1.1.1}$$

The real line $\mathbb{R}$ in the complex plane is the set of solutions z of the equation $\text{Im}(z) = 0$. More generally, let L be a line that contains the point $p \in \mathbb{C}$ and makes an angle θ with the real axis (set $\theta = 0$ if L is parallel to the real axis). If $z \in L$, then $e^{-i\theta}(z - p)$ is real, so $\text{Im}(e^{-i\theta}(z - p)) = 0$. See Figure 1.1.1, p. 2. Conversely, if $e^{-i\theta}(z - p)$ is real, then z lies on L. Multiplying by a positive constant k, and setting $a = ke^{-i\theta}$ and $b = -ke^{-i\theta}p$, we conclude that the line L is the set of solutions to the following equation.

$$\text{Im}(az + b) = 0 \tag{1.1.2}$$

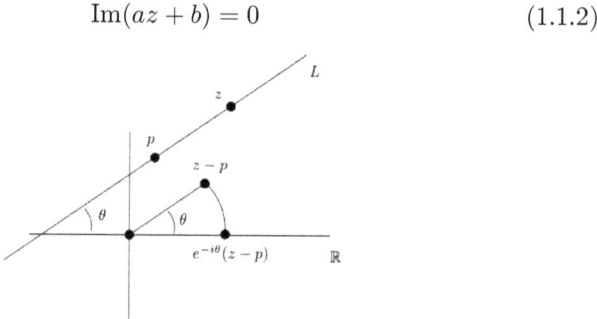

Figure 1.1.1 A line in the complex plane.

Exercises

1. **Solving quadratic equations.** Find all complex solutions of the following equations.

    (a) $z^2 + 3z + 5 = 0$

    (b) $(z - i)(z + i) = 1$

(c) $\dfrac{2z + i}{-z + 3i} = z$

2. **Circles and lines.**

    (a) For a real variable x and a real constant a, *completing the square* refers to rewriting the expression $x^2 - 2ax$ as $(x - a)^2 - a^2$. A complex version of completing the square for a complex variable z and a complex constant a is the following.

$$|z|^2 - 2\operatorname{Re}(za^*) = |z - a|^2 - |a|^2 \qquad (1.1.3)$$

    Write a derivation to justify this. Then use completing the square to find the center and radius of the circle given by the equation $|z|^2 - iz + iz^* - 5 = 0$.

    (b) Write an alternative proof for the general form for the equation of a line (1.1.2), as follows. Let $a = u + iv$, $b = r + is$, $z = x + iy$. Find the equation of the line $\operatorname{Im}(az + b) = 0$ in terms of the real variables x, y and real constants u, v, r, s. Explain why it is necessary that $a \neq 0$.

3. **Complex numbers as 2×2 real matrices.** Let $\mathcal{M}_{\mathbb{C}}$ denote the set of 2×2 matrices of the form $\begin{bmatrix} a & -b \\ b & a \end{bmatrix}$ with $a, b \in \mathbb{R}$. Given a complex number z with Cartesian form $z = a + bi$, let $M(z)$ denote the matrix $\begin{bmatrix} a & -b \\ b & a \end{bmatrix}$ in $\mathcal{M}_{\mathbb{C}}$. Conversely, given a matrix $M \in \mathcal{M}_{\mathbb{C}}$ with top left entry a and bottom left entry b, let $C(M)$ denote the complex number $a + bi$. It is clear that the mappings $z \to M(z)$ and $M \to C(M)$ are inverses to one another, thus establishing a one-to-one correspondence $\mathbb{C} \leftrightarrow \mathcal{M}_{\mathbb{C}}$.

    (a) Show that $\mathcal{M}_{\mathbb{C}}$ is closed under addition and multiplication. That is, suppose that M, N are elements of $\mathcal{M}_{\mathbb{C}}$. Show that $M + N$ and MN are also elements of $\mathcal{M}_{\mathbb{C}}$.

    (b) Show that complex addition and multiplication are "mirrored" in $\mathcal{M}_{\mathbb{C}}$. That is, show that

$$M(z + w) = M(z) + M(w) \qquad (1.1.4)$$
$$M(zw) = M(z)M(w). \qquad (1.1.5)$$

    Significance of this exercise. Matrix algebra provides a framework for theory and applications in almost every area in mathematics. Using the one-to-one correspondence $\mathcal{M}_{\mathbb{C}} \leftrightarrow \mathbb{C}$, it is possible to translate all of complex algebra in terms of matrix operations. We will use this same idea to define and prove the basic properties of quaternion algebra in Section 1.2, p. 5, and we will use correspondences with matrix algebras to prove properties of geometric transformations in Chapter 3, p. 49.

1.2 Quaternions

The quaternions, discovered by William Rowan Hamilton in 1843, were invented to capture the algebra of rotations of 3-dimensional real space, extending the way that the complex numbers capture the algebra of rotations of 2-dimensional real space.

Elements in the set of quaternions $\mathbb{H}$ are in one-to-one correspondence with points in 4-dimensional real space $\mathbb{R}^4$. We will write $r \leftrightarrow (t, x, y, z)$ to denote that the quaternion r corresponds to the 4-tuple (t, x, y, z) of real numbers.

1.2.1 Cartesian form and pure quaternions

The quaternions i, j, k are defined as follows.

$$i \leftrightarrow (0, 1, 0, 0) \tag{1.2.1}$$
$$j \leftrightarrow (0, 0, 1, 0) \tag{1.2.2}$$
$$k \leftrightarrow (0, 0, 0, 1) \tag{1.2.3}$$

The expression $r = a + bi + cj + dk$ is called the **Cartesian form** of the quaternion that corresponds to the vector (a, b, c, d) in $\mathbb{R}^4$. A quaternion of the form $a = a + 0i + 0j + 0k \leftrightarrow (a, 0, 0, 0)$ is called a **scalar** quaternion or a **real** quaternion. A quaternion of the form $xi + yj + zk \leftrightarrow (0, x, y, z)$ is called a **pure** quaternion or an **imaginary** quaternion. For a quaternion $r = a + bi + cj + dk$, we call the real quaternion a the **scalar part** or **real part** of r, and we call the quaternion $xi + yj + zk$ the **vector part** or the **imaginary part** of r. To reflect the natural correspondence of the pure quaternion $xi + yj + zk$ with the vector (x, y, z) in $\mathbb{R}^3$, we will write $\mathbb{R}^3_{\mathbb{H}}$ to denote the space of pure quaternions.

1.2.2 Correspondence with complex matrices

Analogous to the way that the complex numbers $\mathbb{C}$ can be realized as the set $\mathcal{M}_{\mathbb{C}}$ of 2×2 real matrices (see Exercise 1.1.3, p. 3), the quaternions can be realized by a set of 2×2 complex matrices, as follows. Let $\mathcal{M}_{\mathbb{H}}$ denote the set of 2×2 complex matrices of the form $\begin{bmatrix} u & v \\ -v^* & u^* \end{bmatrix}$. Given a quaternion $r = a + bi + cj + dk$, let u, v be the complex numbers $u = a + bi$ and $v = c + di$, and let $M(r)$ denote the 2×2 matrix in $\mathcal{M}_{\mathbb{H}}$ given by

$$M(r) = \begin{bmatrix} u & v \\ -v^* & u^* \end{bmatrix}.$$

Conversely, given a matrix $M \in \mathcal{M}_{\mathbb{H}}$, with top left entry $a + bi$ and top right entry $c + di$, let $Q(M)$ denote the quaternion $r = a + bi + cj + dk$. It is clear that the mappings $r \to M(r)$ and $M \to Q(M)$ are inverses to one another, thus establishing a one-to-one correspondence $\mathbb{H} \leftrightarrow \mathcal{M}_{\mathbb{H}}$.

Proposition 1.2.1 $\mathcal{M}_{\mathbb{H}}$ **is closed under addition and multiplication.** *Let M, N be elements of $\mathcal{M}_{\mathbb{H}}$. Then the sum $M + N$ and the product MN are also elements of $\mathcal{M}_{\mathbb{H}}$.*

Checkpoint 1.2.2 Prove Proposition 1.2.1, p. 6.

1.2.3 Addition and multiplication

By virtue of Proposition 1.2.1, p. 6, we can define addition and multiplication of quaternions r, s as follows.

$$r + s = Q(M(r) + M(s)) \tag{1.2.4}$$
$$rs = Q(M(r)M(s)) \tag{1.2.5}$$

Because matrix algebra has associative and distributive laws, these carry over to quaternions. Note that quaternion multiplication is not commutative! However, for any real quaternion a, we have $M(a) = a\,\mathrm{Id}$, so $M(a)$ commutes with all matrices, and therefore a commutes with all quaternions. To summarize, let q, r, s be quaternions and let a be a real quaternion. We have the following.

$$q(rs) = (qr)s \quad \text{(associative law of multiplication)} \tag{1.2.6}$$
$$q(r + s) = qr + qs \quad \text{(distributive law)} \tag{1.2.7}$$
$$ar = ra \quad \text{(real quaternions commute with all quaternions)} \tag{1.2.8}$$

In practice, it is not necessary to convert quaternions to matrices in order to add and multiply. Quaternion addition and multiplication in Cartesian form is analogous to complex multiplication, using the following basic multiplication rules.

$$i^2 = j^2 = k^2 = -1 \tag{1.2.9}$$
$$ij = -ji = k, \quad jk = -kj = i, \quad ki = -ik = j \tag{1.2.10}$$

Checkpoint 1.2.3 Verify (1.2.9) and (1.2.10).

For $r = a + bi + cj + dk$ and $r' = a' + b'i + c'j + d'k$, we have

$$r + r' = (a + a') + (b + b')i + (c + c')j + (d + d')k. \tag{1.2.11}$$

Multiplication looks like this.

$$\begin{aligned}
rr' &= (a + bi + cj + dk)(a' + b'i + c'j + d'k) \\
&= aa' + bb'i^2 + cc'j^2 + dd'k^2 \\
&\quad + ab'i + ba'i + cd'jk + dc'kj
\end{aligned}$$

$$+\, ac'j + ca'j + bd'ik + db'ki$$
$$+\, ad'k + da'k + bc'ij + cb'ji$$
$$= (aa' - bb' - cc' - dd') \qquad (1.2.12)$$
$$+\, (ab' + ba' + cd' - dc')i$$
$$+\, (ac' + ca' - bd' + db')j$$
$$+\, (ad' + da' + bc' - cb')k$$

If u, v are pure quaternions, (1.2.12) can be written more compactly in terms of the dot and cross products for vectors in $\mathbb{R}^3$.

$$uv = -(u \cdot v) + u \times v \quad \text{(for pure quaternions } u, v) \qquad (1.2.13)$$

Checkpoint 1.2.4 Verify (1.2.13).

1.2.4 Conjugate, modulus, and polar form

The **conjugate** of a quaternion $r = a + bi + cj + dk$ is $r^* = a - bi - cj - dk$, and the **modulus** of r is $|r| = \sqrt{a^2 + b^2 + c^2 + d^2}$. The **unit quaternions**, denoted $U(\mathbb{H})$, is the set of quaternions with modulus 1.[1] Analogous to complex numbers, a quaternion r can be expressed in **polar form**

$$r = |r|(\cos\theta + u\sin\theta) \qquad (1.2.14)$$

where u is a pure unit quaternion and θ is a real number.

Checkpoint 1.2.5

1. Show that the following construction produces a polar form for a nonzero quaternion r. Let $r' = \frac{r}{|r|} = a' + b'i + c'j + d'k$. If $|a'| < 1$, let $u = \frac{1}{\sqrt{1-(a')^2}}(b'i + c'j + d'k)$, and let $\theta = \arccos a'$.

2. Fill in the remaining details on polar form for quaternions. What happens if $r = 0$? What happens if $|a'| = 1$?

3. Are u, θ uniquely determined by r? If not, describe the possible choices for u, θ.

Continuing the analogy with complex numbers, we have the following, for all quaternions r, s.

$$(rs)^* = s^* r^* \qquad (1.2.15)$$
$$|r|^2 = rr^* = r^* r \qquad (1.2.16)$$
$$|rs| = |r||s|. \qquad (1.2.17)$$

---

[1]The set of unit quaternions $U(\mathbb{H})$ is in one-to-one correspondence with the 3-sphere $S^3 = \{(t, x, y, z) \in \mathbb{R}^4 : t^2 + x^2 + y^2 + z^2 = 1\}$. This is analogous to the set of norm 1 complex numbers that is in one-to-one correspondence with the 1-sphere $S^1 = \{(x, y) \in \mathbb{R}^2 : x^2 + y^2 = 1\}$.

Checkpoint 1.2.6 Verify the three equations above.

Hint.   Work in $\mathcal{M}_{\mathbb{H}}$. Start by checking that $(M(r^*)) = (M(r))^{\dagger}$, where $\dagger$ denotes the conjugate transpose of a matrix. Alternatively, write r, s in polar form and use (1.2.13).

1.2.5 Quaternions as rotations of $\mathbb{R}^3_{\mathbb{H}}$

Let r be a unit quaternion and let v be a pure quaternion. Let $R_r(v)$ denote the quaternion $R_r(v) = rvr^*$. It is easy to check that $(R_r(v))^* = -R_r(v)$. From this we conclude that rvr^* is a pure quaternion.

Checkpoint 1.2.7 Explain how "we conclude" that $R_r(v)$ is pure when r is a unit quaternion and v is a pure quaternion.

It is easy to see that R_r is a linear map from the real vector space of unit quaternions to itself. That means that the following properties hold for all pure quaternions v, w and all real scalars α.

$$R_r(v + w) = R_r(v) + R_r(w) \tag{1.2.18}$$
$$R_r(\alpha v) = \alpha R_r(v) \tag{1.2.19}$$

Checkpoint 1.2.8 Show the details to prove that R_r is linear.

We conclude with the main result of this section that shows how rotations of 3-dimensional real space are encoded in the algebra of quaternions.

Proposition 1.2.9   Quaternions and rotations of $\mathbb{R}^3_{\mathbb{H}}$.   *Let $r = \cos\theta + u\sin\theta$ be a unit quaternion in polar form, and let R_r be the linear transformation of the space of pure quaternions given by $v \to rvr^*$. The action of R_r is a rotation by 2θ radians about the axis given by the unit vector u.*

1.2.6 Exercises

1.   Prove Proposition 1.2.9, p. 8 using the following outline. Let $r = \cos\theta + u\sin\theta$ be a polar form for a unit quaternion r.

   (a) Show that $R_r(u) = u$.

   (b) Let v be any pure unit quaternion orthogonal to u, and let $w = u \times v$, so that the triple u, v, w forms a right-handed coordinate system for $\mathbb{R}^3$. Show that

$$R_r(v) = \cos(2\theta)v + \sin(2\theta)w \tag{1.2.20}$$

   (use equation (1.2.13)) and then explain how this proves the Proposition.

   Hint.   In deriving equation (1.2.20), you will obtain expressions $uv - vu$ and uvu. Use equation (1.2.13) to show that $uv - vu = 2w$ and $uvu = v$.

Show that the quaternion on the right side of (1.2.20) has norm 1. Finally, use the fact that

$$a \cdot b = |a||b| \cos t$$

for real vectors a, b that make an angle t at the origin to determine the angle made by $v, R_r(v)$.

2.   Show that the following hold for all $r, s \in U(\mathbb{H})$.

   (a) $R_r \circ R_s = R_{rs}$

   (b) $(R_r)^{-1} = R_{r^*}$

1.3 Stereographic projection

1.3.1 Stereographic projection $S^1 \to \hat{\mathbb{R}}$

Let $S^1 = \{(x, y) \in \mathbb{R}^2 : x^2 + y^2 = 1\}$ denote the unit circle in the x, y-plane. Let $N = (0, 1)$ denote the "north pole" (that is, the point at the "top" of the unit circle). Given a point $P = (x, y) \neq N$ on the unit circle, let $s(P)$ denote the intersection of the line $\overline{NP}$ with the x-axis. See Figure 1.3.1, p. 10. The map $s \colon S^1 \setminus \{N\} \to \mathbb{R}$ given by this rule is called **stereographic projection**. Using similar triangles, it is easy to see that $s(x, y) = \frac{x}{1-y}$.

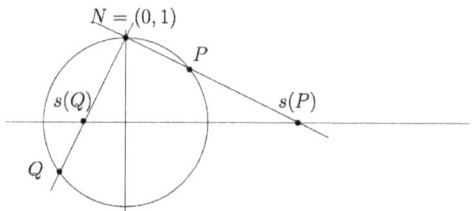

Figure 1.3.1 Stereographic projection

Checkpoint 1.3.2 Draw the relevant similar triangles and verify the formula $s(x, y) = \frac{x}{1-y}$.

We extend stereographic projection to the entire unit circle as follows. We call the set

$$\hat{\mathbb{R}} = \mathbb{R} \cup \{\infty\} \tag{1.3.1}$$

the **extended real numbers**, where "∞" is an element that is not a real number. Now we define stereographic projection $s \colon S^1 \to \hat{\mathbb{R}}$ by

$$s(x, y) = \begin{cases} \frac{x}{1-y} & y \neq 1 \\ \infty & y = 1. \end{cases} \tag{1.3.2}$$

1.3.2 Stereographic projection $S^2 \to \hat{\mathbb{C}}$

The definitions in the previous subsection extend naturally to higher dimensions. Here are the details for the main case of interest.

Let $S^2 = \{(a, b, c) \in \mathbb{R}^3 : a^2 + b^2 + c^2 = 1\}$ denote the unit sphere in $\mathbb{R}^3$. Let $N = (0, 0, 1)$ denote the "north pole" (that is, the point at the "top" of the sphere, where the positive z-axis is "up"). Given a point $P = (a, b, c) \neq N$ on the unit sphere, let $s(P)$ denote the intersection of the line $\overline{NP}$ with the x, y-plane, which we identify with the complex plane $\mathbb{C}$. See Figure 1.3.3, p. 11. The map $s \colon S^2 \setminus \{N\} \to \mathbb{C}$ given by this rule is called **stereographic projection**. Using similar triangles, it is easy to see that $s(a, b, c) = \frac{a+ib}{1-c}$.

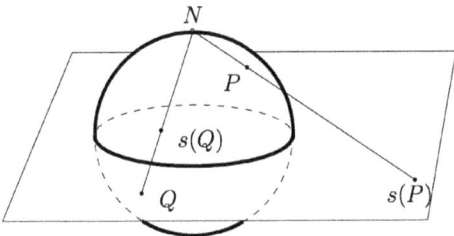

Figure 1.3.3 Stereographic projection

We extend stereographic projection to the entire unit sphere as follows. We call the set

$$\hat{\mathbb{C}} = \mathbb{C} \cup \{\infty\} \tag{1.3.3}$$

the **extended complex numbers**, where "∞" is an element that is not a complex number. Now we define stereographic projection $s \colon S^2 \to \hat{\mathbb{C}}$ by

$$s(a, b, c) = \begin{cases} \frac{a+ib}{1-c} & c \neq 1 \\ \infty & c = 1. \end{cases} \tag{1.3.4}$$

1.3.3 Conjugate Transformations

Let $\mu \colon X \to Y$ be a bijective map. We say that maps and $f \colon X \to X$ and $g \colon Y \to Y$ are **conjugate transformations** (with respect to the bijection μ) if $f = \mu^{-1} \circ g \circ \mu$. Exercise Group 1.3.4.3–6, p. 11 demonstrate examples of this definition for which μ is stereographic projection.

Checkpoint 1.3.4 Show that $f = \mu^{-1} \circ g \circ \mu$ if and only if $\mu \circ f = g \circ \mu$.

1.3.4 Exercises

Formulas for inverse stereographic projection. It is intuitively clear that stereographic projection is a bijection. Make this rigorous by finding a formula for the inverse.

1. For $s \colon S^1 \to \hat{\mathbb{R}}$, find a formula for $s^{-1} \colon \hat{\mathbb{R}} \to S^1$. Find $s^{-1}(3)$.
   (See *Solutions to Exercises* in Appendix B, p. 103.)

2. For $s \colon S^2 \to \hat{\mathbb{C}}$, find a formula for $s^{-1} \colon \hat{\mathbb{C}} \to S^2$. Find $s^{-1}(3+i)$.
   (See *Solutions to Exercises* in Appendix B, p. 103.)

Transformations that are conjugate via stereographic projection. *Suggestion:* Use Checkpoint 1.3.4, p. 11 for the exercises that follow.

3.   Let $f\colon S^1 \to S^1$ and $g\colon \hat{\mathbb{R}} \to \hat{\mathbb{R}}$ be given by

$$f(x,y) = (x, -y)$$

$$g(x) = \begin{cases} 1/x & x \neq 0, \infty \\ \infty & x = 0 \\ 0 & x = \infty. \end{cases}$$

Show that f, g are conjugate transformations with respect to stereographic projection.

4.   Let $R_{Z,\theta}\colon S^2 \to S^2$ and $T_{Z,\theta}\colon \hat{\mathbb{C}} \to \hat{\mathbb{C}}$ be given by

$$R_{Z,\theta}(a,b,c) = (a\cos\theta - b\sin\theta, a\sin\theta + b\cos\theta, c)$$

$$T_{Z,\theta}(z) = \begin{cases} e^{i\theta} z & z \neq \infty \\ \infty & z = \infty. \end{cases}$$

Show that $R_{Z,\theta}, T_{Z,\theta}$ are conjugate transformations with respect to stereographic projection.

5.   Let $R_{X,\pi}\colon S^2 \to S^2$ and $T_{X,\pi}\colon \hat{\mathbb{C}} \to \hat{\mathbb{C}}$ be given by

$$R_{X,\pi}(a,b,c) = (a, -b, -c)$$

$$T_{X,\pi}(z) = \begin{cases} 1/z & z \neq 0, \infty \\ \infty & z = 0 \\ 0 & z = \infty. \end{cases}$$

Show that $R_{X,\pi}, T_{X,\pi}$ are conjugate transformations with respect to stereographic projection.

6.   Let $R_{X,\pi/2}\colon S^2 \to S^2$ and $T_{X,\pi/2}\colon \hat{\mathbb{C}} \to \hat{\mathbb{C}}$ be given by

$$R_{X,\pi/2}(a,b,c) = (a, -c, b)$$

$$T_{X,\pi/2}(z) = \begin{cases} \frac{z+i}{iz+1} & z \neq i, \infty \\ \infty & z = i \\ -i & z = \infty. \end{cases}$$

Show that $R_{X,\pi/2}, T_{X,\pi/2}$ are conjugate transformations with respect to stereographic projection.

7. Projections of endpoints of diameters.   Show that $s(a,b,c)(s(-a,-b,-c))^* = -1$ for any point (a,b,c) in S^2 with $|c| \neq 1$. Conversely, suppose that $zw^* = -1$ for some $z, w \in \mathbb{C}$. Show that $s^{-1}(z) = -s^{-1}(w)$.

1.4 Equivalence relations

1.4.1 Definitions

A relation on a set X is a subset of $X \times X$. Given a relation $R \subseteq X \times X$, we write $x \sim_R y$, or just $x \sim y$ if R is understood by context, to denote that $(x, y) \in R$. A relation is **reflexive** if $x \sim x$ for every x in X. A relation is **symmetric** if $x \sim y$ implies $y \sim x$. A relation is **transitive** if $x \sim y$ and $y \sim z$ together imply that $x \sim z$. A relation is called an **equivalence relation** if it is reflexive, symmetric, and transitive. Given an equivalence relation on X and an element $x \in X$, we write $[x]$ to denote the set

$$[x] = \{y \in X : x \sim y\}, \tag{1.4.1}$$

called the **equivalence class** of the element x. The set of equivalence classes is denoted $X/\sim$, that is,

$$X/\sim = \{[x] : x \in X\}. \tag{1.4.2}$$

A **partition** of a set X is a collection of nonempty disjoint sets whose union is X.

1.4.2 The integers modulo an integer n

Let n be a positive integer. Let $\sim_n$ be the relation on the integers $\mathbb{Z}$ given by

$$x \sim_n y \Leftrightarrow n | (x - y)$$

(recall that the symbols "$a|b$" for integers a, b, pronounced "a divides b", means $b = ka$ for some integer k). It is easy to show that $\sim_n$ is an equivalence relation, and that the equivalence classes are precisely the set

$$\mathbb{Z}/\sim_n = \{[0], [1], [2], \ldots, [n-1]\}.$$

This set of equivalence classes is fundamental and pervasive in mathematics. Using standard notation, we write

$$x = y \pmod{n}$$

(pronounced "x is equivalent to y mod n") to denote "$x \sim_n y$". In this text, we will follow the common practice of using $\mathbb{Z}_n$ to denote $\mathbb{Z}/\sim_n$.

Checkpoint 1.4.1

1. Verify that the relation $\sim_n$ is indeed an equivalence relation.

2. Verify that the equivalence classes of the equivalence relation $\sim_n$ are indeed $\{[0], [1], [2], \ldots, [n-1]\}$. Hint: Use the **division algorithm**,

which says that for any $x \in \mathbb{Z}$, there are unique integers q, r, with r in the range $0 \leq r \leq n - 1$, such that $x = qn + r$.

3. Draw a sketch that shows how $\mathbb{Z}$ is partitioned by the mod n equivalence classes.

1.4.3 Facts

Fact 1.4.2 Equivalence relations and partitions. *Let X be a set. Equivalence relations on X and partitions of X are in one-to-one correspondence, as follows. Given an equivalence relation $\sim$ on X, the collection*

$$X/\sim = \{[x] : x \in X\}$$

is a partition of X. Conversely, given a partition $\mathcal{P}$ of X, the relation $\sim_\mathcal{P}$ defined by

$$x \sim_\mathcal{P} y \Leftrightarrow x, y \text{ lie in the same element of } \mathcal{P}$$

is an equivalence relation. These correspondences are inverse to one another. That is, $\sim = \sim_{(X/\sim)}$ and $X/(\sim_\mathcal{P}) = \mathcal{P}$.

Checkpoint 1.4.3 A key component of Fact 1.4.2, p. 14 is that equivalence classes are disjoint. Let $\sim$ be an equivalence relation on a set X, and let $x, y \in X$. Show that either $[x] = [y]$ or $[x] \cap [y] = \emptyset$.

Fact 1.4.4 Construction of functions on sets of equivalence classes. *Let $\sim$ be an equivalence relation on a set X, let $\pi \colon X \to X/\sim$ denote the map given by $x \to [x]$, and let $f \colon X \to Y$ be a function. There exists a map $\overline{f} \colon X/\sim \to Y$ such that $(\overline{f} \circ \pi)(x) = f(x)$ for all $x \in X$ if and only if f is constant on equivalence classes (that is, if and only if $[x \sim y \Rightarrow f(x) = f(y)]$).*

Note on terminology: When a function f is constant on equivalence classes, we say that the associated function $\overline{f}$ is **well-defined**.

Fact 1.4.5 Construction of equivalence relations and partitions from functions. *Given a function $f \colon X \to Y$, there is a natural equivalence relation $\sim_f$ on X given by*

$$x \sim_f y \Leftrightarrow f(x) = f(y).$$

The corresponding set of equivalence classes is $X/\sim_f = \{f^{-1}(y) : y \in f(X)\}$. Furthermore, the function $X/\sim_f \to f(X)$ given by $[x] \to f(x)$ is a one-to-one correspondence.

1.4.4 Exercises

1. **The rational numbers.** Let X be the set

$$X = \mathbb{Z} \times (\mathbb{Z} \setminus \{0\}) = \{(m, n) \colon m, n \in \mathbb{Z}, n \neq 0\}$$

and let $\mathbb{Q}$ denote the set of rational numbers

$$\mathbb{Q} = \left\{ \frac{m}{n} \colon m, n \in \mathbb{Z}, n \neq 0 \right\}.$$

Define the relation $\sim$ on X by

$$(m, n) \sim (m', n') \Leftrightarrow mn' - m'n = 0.$$

Let $f \colon X \to \mathbb{Z}$ be given by $f(m, n) = m^2 + n^2$ and let $g \colon X \to \mathbb{Q}$ be given by $g(m, n) = \frac{m}{n}$.

(a) Show that $\sim$ is reflexive, symmetric, and transitive.

(b) Draw a sketch of X showing the partition $\{[(m, n)] \colon (m, n) \in X\}$.

(c) Is the map $\overline{f} \colon X/\sim \to \mathbb{Z}$ given by $\overline{f}([(m, n)]) = f(m, n)$ well-defined? Explain.

(d) Draw a sketch of X that shows the equivalence classes $X/\sim_f$.

(e) Is the map $\overline{g} \colon X/\sim \to \mathbb{Q}$ given by $\overline{g}([(m, n)]) = g(m, n)$ well-defined? Explain.

(f) Draw a sketch of X that shows the equivalence classes $X/\sim_g$.

(g) Explain why $\sim$ and $\sim_g$ are the same equivalence relation on X. Explain why $X/\sim$ is in one-to-one correspondence with the rational numbers $\mathbb{Q}$.

The integers modulo n. Let n be a positive integer.

2. Let ω be the complex number $\omega = e^{2\pi i/n}$, and let $f \colon \mathbb{Z} \to \mathbb{C}$ be given by $m \to \omega^m$. Show that the equivalence relation $\sim_f$ given by Fact 1.4.5, p. 14 is the same as $\sim_n$.

3. Show that the operation of addition on $\mathbb{Z}_n$ given by

$$[x] + [y] := [x + y]$$

is well-defined. This means showing that if $[x] = [x']$ and $[y] = [y']$, then $[x + y] = [x' + y']$.

4.   Show that the operation of multiplication on $\mathbb{Z}_n$ given by

$$[x] \cdot [y] := [xy]$$

is well-defined.

5.   **Alternative construction of $\mathbb{Z}_n$.** Another standard definition
of the set $\mathbb{Z}_n$, together with its operations of addition and multiplication, is the following. Given an integer a, we write $a \text{ MOD } n$
to denote the **remainder** obtained when dividing a by n (the
integer $a \text{ MOD } n$ is the same as the integer r in the statement of
the division algorithm given in Checkpoint 1.4.1, p. 13). Now define
$\mathbb{Z}_n$ to be the set

$$\mathbb{Z}_n = \{0, 1, 2, \ldots, n-1\},$$

define the addition operation $+_n$ on $\mathbb{Z}_n$ by

$$x +_n y = (x+y) \text{ MOD } n,$$

and define the multiplication operation $\cdot_n$ on $\mathbb{Z}_n$ by

$$x \cdot_n y = (xy) \text{ MOD } n.$$

Show that this version of $\mathbb{Z}_n$ is equivalent to the version developed
in Exercise 1.4.4.3, p. 15 and Exercise 1.4.4.4, p. 16.

1.5 More preliminary topics

1.5.1 A useful tool: commutative diagrams

A **directed graph** (or *digraph*) is a set V of *vertices* and a set $E \subset V \times V$ of *directed edges*. We draw pictures of digraphs by drawing an arrow pointing from a vertex v to a vertex w whenever $(v, w) \in E$. See Figure 1.5.1, p. 17.

A **path** in a directed graph is a sequence of vertices $v_0, v_1, \ldots, v_n$ such that $(v_{i-1}, v_i) \in E$ for $1 \leq i \leq n$. The vertex v_0 is called the **initial vertex** and v_n is called the **final vertex** of the path $v_0, v_1, \ldots, v_n$.

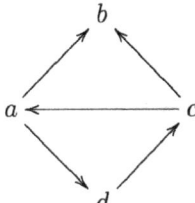

Figure 1.5.1 Example of a directed graph with vertex set $V = \{a, b, c, d\}$ and edge set $E = \{(a, b), (c, b), (c, a), (a, d), (d, c)\}$. The vertex sequences c, b and c, a, b are both paths from c to b.

A **commutative diagram** is a directed graph with two properties.

1. Vertices are labeled by sets and directed edges are labeled by functions between those sets. That is, the directed edge $f = (X, Y)$ denotes a function $f \colon X \to Y$.

2. Whenever there are two paths from an initial vertex X to a final vertex Y, the composition of functions along one path is equal to the composition of functions along the other path. That is, if $X_0, X_1, \ldots, X_n$ is a path with edges $f_i \colon X_{i-1} \to X_i$ for $1 \leq i \leq n$ and $X_0 = Y_0, Y_1, Y_2, \ldots, Y_m = X_n$ is a path with edges $g_i \colon Y_{i-1} \to Y_i$ for $1 \leq i \leq m$, then

$$f_n \circ f_{n-1} \circ \cdots \circ f_1 = g_m \circ g_{m-1} \circ \cdots \circ g_1.$$

Figure 1.5.2, p. 18 shows a commutative diagram that illustrates the definition of conjugate transformations. Figure 1.5.3, p. 18 shows a commutative diagram that goes with Fact 1.4.4, p. 14.

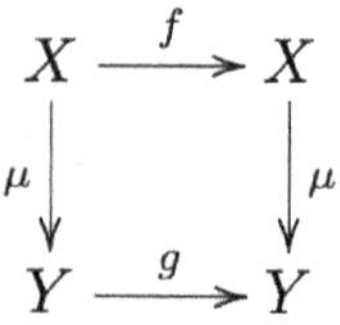

$$
\begin{array}{ccc}
X & \xrightarrow{\ f\ } & X \\
\downarrow{\scriptstyle \mu} & & \downarrow{\scriptstyle \mu} \\
Y & \xrightarrow{\ g\ } & Y
\end{array}
$$

Figure 1.5.2 A commutative diagram illustrating the definition of conjugate transformations f, g given in Exercise Group 1.3.4.3–6, p. 11.

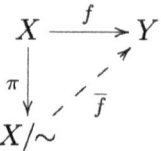

Figure 1.5.3 A commutative diagram showing the relationship $\overline{f} \circ \pi = f$ in Fact 1.4.4, p. 14.

1.5.2 Exercises

1. Let r be a pure, unit quaternion. Use (1.2.13) to show that the map $\mathbb{R}^3_{\mathbb{H}} \to \mathbb{R}^3_{\mathbb{H}}$ given by $u \to rur$ is the reflection across the plane normal to r. That is, show that $rur = u - 2(u \cdot r)r$. See Figure 1.5.4, p. 18.

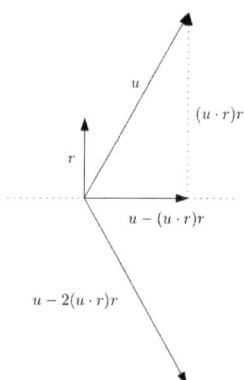

Figure 1.5.4 The reflection of $u \in \mathbb{R}^3_{\mathbb{H}}$ across the plane normal to $r \in \mathbb{R}^3_{\mathbb{H}}$.

2. **Commutative diagram examples.**

    (a) Draw a commutative diagram that illustrates the results of Exercise 1.3.4.5, p. 12.

(b) The **distributive law** for $\mathbb{Z}_n$ says that

$$[x]\,([y] + [z]) = [x][y] + [x][z]$$

for all $[x], [y], [z] \in \mathbb{Z}_n$. Label the maps in the commutative diagram below (Figure 1.5.5, p. 19) to express the distributive law.

$$
\begin{array}{ccc}
\mathbb{Z}_n \times \mathbb{Z}_n & \longrightarrow & \mathbb{Z}_n \\
\downarrow & & \downarrow \\
\mathbb{Z}_n \times \mathbb{Z}_n & \longrightarrow & \mathbb{Z}_n
\end{array}
$$

Figure 1.5.5

Chapter 2

Groups

2.1 Examples of groups

Groups are one of the most basic algebraic objects, yet have structure rich enough to be widely useful in all branches of mathematics and its applications. A group is a set G with a binary operation $G \times G \to G$ that has a short list of specific properties. Before we give the complete definition of a group in the next section (see Definition 2.2.1, p. 28), this section introduces examples of some important and useful groups.

2.1.1 Permutations

A **permutation** of a set X is a bijection from X to itself, that is, a function that is both one-to-one and onto. Given two permutations α, β of a set X, we write $\alpha\beta$ to denote the composition of functions $\alpha \circ \beta$.

Definition 2.1.1 Let X be a set and let $\mathrm{Perm}(X)$ denote the set of all permutations of X. The *group of permutations of X* is the set $G = \mathrm{Perm}(X)$ together with the binary operation $G \times G \to G$ given by function composition, that is, $(\alpha, \beta) \to \alpha \circ \beta$. For the special case $X = \{1, 2, \ldots, n\}$ for some integer $n \geq 1$, the group $\mathrm{Perm}(X)$ is called the **symmetric group**, and is denoted S_n. $\Diamond$

   Notation: We will denote the element σ in S_n using the symbols $[\sigma(1), \sigma(2), \ldots, \sigma(n)]$, that is, the list of values of σ, separated by commas and enclosed in square brackets.[1] For example, we write $[3, 1, 2]$ to denote the permutation $\sigma \colon \{1, 2, 3\} \to \{1, 2, 3\}$ given by

$$\sigma(1) = 3, \ \sigma(2) = 1, \ \sigma(3) = 2.$$

Checkpoint 2.1.2 Let $\sigma = [3, 1, 2]$ and let $\tau = [2, 1, 3]$. Find $\sigma\tau$, $\tau\sigma$, and $\sigma^2 = \sigma\sigma$.

   (See *Solutions to Exercises* in Appendix B, p. 103.)

2.1.2 Symmetries of regular polygons

Informally and intuitively, we say that regular polygons have rotational and mirror symmetries. Specifically, the rotational symmetries are rotations about the center O of the polygon, clockwise or counterclockwise, by some angle $\angle POP'$, where P, P' are any two vertices. The mirror symmetries of the polygon are reflections across lines of the form $\overline{OP}$ or $\overline{OM}$, where

---

[1] Writing $[\sigma(1), \sigma(2), \ldots, \sigma(n)]$ to denote $\sigma \in S_n$ is a compact replacement for the standard ``input-output'' 2-row notation

$$\begin{bmatrix} 1 & 2 & \cdots & n \\ \sigma(1) & \sigma(2) & \cdots & \sigma(n) \end{bmatrix}$$

that is used in most introductory textbooks.

P is any vertex and M is the midpoint of any edge of the polygon. See Figure 2.1.3, p. 23.

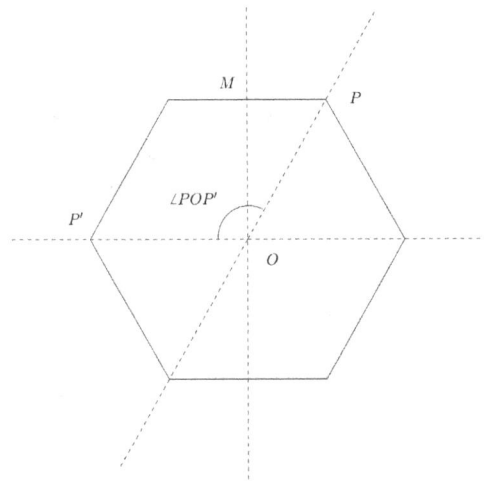

Figure 2.1.3 Symmetries of a regular n-gon

Here are some standard notations for rotations and reflections in the plane. See Figure 2.1.4, p. 23.

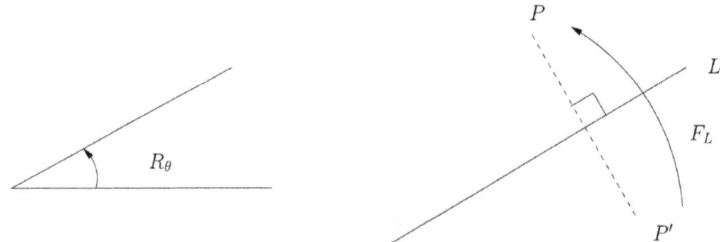

Figure 2.1.4 Rotations and reflections in the plane

Rotations in the plane.

Fix a center point O. We write R_θ to denote the rotation by angle θ about the point O. We observe the usual convention that positive values of θ denote counterclockwise rotations and negative values of θ denote clockwise rotations.[2]

---

[2]Angle units can be radians, degrees, revolutions, or whatever is most convenient. It is the responsibility of the user to be clear.

Reflections in the plane.

We write F_L to denote the reflection across the line L. This means that $P' = F_L(P)$ if and only if $\overline{PP'} \perp L$ and the distance from P to L is the same as the distance from P' to L.

Given symmetries A, B, we write AB to denote the composition $A \circ B$. For example, for the symmetries of the equilateral triangle, with angles in degrees, and with $L = \overline{OP}$ for some vertex P, we have $R_{240}R_{120} = R_0$ and $F_L R_{120} = R_{-120} F_L$.

Definition 2.1.5 The **dihedral group**, denoted D_n , is the set of rotation and reflection symmetries of the regular n-gon together with the binary operation of function composition. ◇

Checkpoint 2.1.6 Let X be the square centered at the origin in the x, y-plane with vertices at $(\pm 1, \pm 1)$. The square X has lines of symmetry H, V, D, D' (horizontal, vertical, diagonal, and another diagonal) where H, V denote the x, y axes, respectively, and D, D' denote the lines $y = -x, y = x$, respectively. See Figure 2.1.7, p. 24.

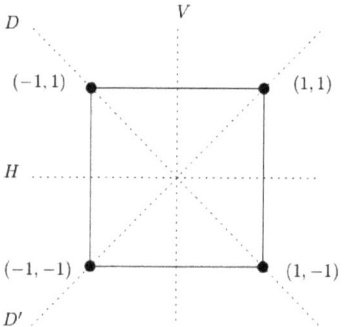

Figure 2.1.7 Lines of symmetry for the square.

The symmetries of the square X are

$$D_4 = \{R_0, R_{1/4}, R_{1/2}, R_{3/4}, F_H, F_V, F_D, F_{D'}\}$$

where the rotation angles units are revolutions. Find the following.

1. $R_{1/4}R_{1/2}$

2. $R_{1/4}F_H$

3. $F_H R_{1/4}$

4. $F_H F_D$

5. $F_D F_H$

6. $(F_D R_{1/2})^2 = F_D R_{1/2} F_D R_{1/2}$

7. $(F_D R_{1/2})^3$

(See *Solutions to Exercises* in Appendix B, p. 103.)

2.1.3 The norm 1 complex numbers

Definition 2.1.8 The **circle group**, denoted S^1, is the set

$$S^1 = \{z \in \mathbb{C} : |z| = 1\}$$

of norm 1 complex numbers together with the binary operation $S^1 \times S^1 \to S^1$ given by complex multiplication, that is, $(z, w) \to zw$.   ◊

Checkpoint 2.1.9 Show that if z, w are elements of S^1, then their product zw is also in S^1.

2.1.4 The n-th roots of unity

Let $n \geq 1$ be an integer. The set

$$C_n = \{z \in \mathbb{C} : z^n = 1\}$$

is called the set of **(complex) n-th roots of unity**.

Checkpoint 2.1.10

1. Let $\omega = e^{i2\pi/n}$. Show that ω^k is in C_n for all integers k.

2. Show that, if z is an element of C_n, then $z = \omega^k$ for some integer k.

3. Show that the set C_n consists of precisely the n elements

$$\{\omega^0, \omega^1, \omega^2, \dots, \omega^{n-1}\}.$$

Definition 2.1.11 The set $C_n = \{\omega^0, \omega^1, \omega^2, \dots, \omega^{n-1}\}$, together with the operation of complex multiplication, is called the **group of n-th roots of unity**.   ◊

2.1.5 Integers

Definition 2.1.12 The set $\mathbb{Z}$ of integers, together with the operation of addition, is called the **group of integers**. Similarly, the set $\mathbb{Z}_n$ of integers modulo n (where n is some integer $n \geq 1$), together with the operation of addition modulo n, is called the **group of integers mod n**.   ◊

2.1.6 Invertible matrices

Let $n \geq 1$ be an integer. We write $GL(n, \mathbb{R})$ to denote the set of $n \times n$ invertible matrices with real entries. We write $GL(n, \mathbb{C})$ to denote the set of $n \times n$ invertible matrices with complex entries.

Definition 2.1.13 The set $GL(n, \mathbb{R})$ (respectively, $GL(n, \mathbb{C})$), together with the binary operation of matrix multiplication, is called the **group of** $n \times n$ **real (respectively, complex) invertible matrices**, or also the **general linear group**. ◊

2.1.7 Nonzero elements in a field

Let $\mathbb{F}$ be a field, such as the rational numbers $\mathbb{Q}$, the real numbers $\mathbb{R}$, or the complex numbers $\mathbb{C}$. We write $\mathbb{F}^*$ to denote the set of nonzero elements in $\mathbb{F}$.

Definition 2.1.14 Let $\mathbb{F}$ be a field. The set $\mathbb{F}^*$, together with the binary operation of multiplication, is called the **group of nonzero elements** in the field $\mathbb{F}$. ◊

2.1.8 Unit quaternions

Definition 2.1.15 The set $U(\mathbb{H})$ of quaternions of norm 1 (defined in Subsection 1.2.4, p. 7), together with the binary operation of quaternion multiplication, is called the **group of unit quaternions**. ◊

2.1.9 Exercises

1. **Matrices for the dihedral group** D_4. Let H denote the x-axis in the x, y-plane. The map $F_H : \mathbb{R}^2 \to \mathbb{R}^2$ is a linear map whose matrix is $\begin{bmatrix} 1 & 0 \\ 0 & -1 \end{bmatrix}$. The map $R_{1/4} : \mathbb{R}^2 \to \mathbb{R}^2$ is a linear map whose matrix is $\begin{bmatrix} 0 & -1 \\ 1 & 0 \end{bmatrix}$. Find the matrices for the remaining elements of the dihedral group D_4 as specified in Checkpoint 2.1.6, p. 24.

2. **Complex number operations for the dihedral group** D_4. Let H denote the real line $\mathbb{R}$ in the complex plane $\mathbb{C}$. The map $F_H : \mathbb{C} \to \mathbb{C}$ is complex conjugation $z \to z^*$. The map $R_{1/4} : \mathbb{C} \to \mathbb{C}$ is the map $z \to e^{i\pi/2}z = iz$. Find the maps $\mathbb{C} \to \mathbb{C}$ for the remaining elements of the dihedral group D_4 as specified in Checkpoint 2.1.6, p. 24.

3. Recall that a binary operation $(x, y) \to x*y$ is *commutative* if $x*y = y*x$ for all possible values of x, y.

    (a) Which of the group operations in the examples in this section are commutative? Which are not?

(b) Show that S_n is *not* commutative for $n > 2$.

4. One of the properties of a group is the existence of an *identity* element. This is a group element e with the property that $eg = ge = g$ for every g in G. Find an identity element for each of the groups in the examples in this section.

5. One of the properties of a group is the existence of an *inverse* element for every element in the group. This means that for every g in a group G, there is an element h with the property that $gh = hg = e$, where e is the identity element of the group. Find inverses for the following list of group elements.

  (a) $[4, 2, 1, 3]$ in S_4

  (b) R_{120} in D_6 (where 120 is in degrees)

  (c) $\frac{1}{\sqrt{2}}(-1 + i)$ in S^1

  (d) 7 in $\mathbb{Z}$

  (e) 7 in $\mathbb{Z}_9$

  (f) $\begin{bmatrix} 1 & 2 \\ 2 & 1 \end{bmatrix}$ in $GL(2, \mathbb{R})$

  (g) $r = a + bi + cj + dk$ in $U(\mathbb{H})$

2.2 Definition of a group

We will use the notation $*\colon S \times S \to S$ to denote a binary operation on a set S that sends the pair (x, y) to $x * y$. Recall that a binary operation $*$ is *associative* means that $x * (y * z) = (x * y) * z$ for all $x, y, z \in S$.

Definition 2.2.1 Group. A **group** is a set G, together with a binary operation $*\colon G \times G \to G$ with the following properties.

- The operation $*$ is associative.

- There exists an element e in G, called an **identity** element, such that $e * g = g * e = g$ for all $g \in G$.

- For every $g \in G$, there exists an element $h \in G$, called an **inverse** element for g, such that $g * h = h * g = e$.

$\Diamond$

Proposition 2.2.2 Immediate consequences of the definition of group. *Let G be a group. The element e in the second property of Definition 2.2.1, p. 28 is unique. Given $g \in G$, the element h in the third property of Definition 2.2.1, p. 28 is unique.*

Definition 2.2.3 Multiplicative notation. Let G be a group. By Proposition 2.2.2, p. 28, we may speak of an identity element as *the* identity element for G. Given $g \in G$, we may refer to an inverse element for g as *the* inverse of g, and we write g^{-1} to denote this element. In practice, we often omit the operator $*$, and simply write gh to denote $g * h$. We adopt the convention that g^0 is the identity element. For $k \geq 1$, we write g^k to denote $\underbrace{g * g * \cdots * g}_{k \text{ factors}}$ and we write g^{-k} to denote $\left(g^k\right)^{-1}$. This set of notational conventions is called **multiplicative notation.** $\Diamond$

Definition 2.2.4 Abelian group, additive notation. In general, group operations are not commutative (see Exercise 2.1.9.3, p. 26). A group with a commutative operation is called **Abelian.** For some Abelian groups, such as the group of integers, the group operation is called *addition*, and we write $a + b$ instead of using the multiplicative notation $a * b$. We write 0 to denote the identity element, we write $-a$ to denote the inverse of a, and we write ka to denote $\underbrace{a + a + \cdots + a}_{k \text{ summands}}$ for positive integers k. This set of notational conventions is called **additive notation.** $\Diamond$

Definition 2.2.5 Order of a group. The number of elements in a finite group is called the **order** of the group. A group with infinitely many elements is said to be of *infinite order*. We write $|G|$ to denote the order of the group G. $\Diamond$

Definition 2.2.6 The trivial group. A group with a single element (which is necessarily the identity element) is called a **trivial group**. In multiplicative notation, one might write $\{1\}$, and in additive notation, one might write $\{0\}$, to denote a trivial group. ◇

Exercises

1. **Uniqueness of the identity element.** Let G be a group. Suppose that e, e' both satisfy the second property of the Definition 2.2.1, p. 28, that is, suppose $e * x = x * e = e' * x = x * e' = x$ for all $x \in G$. Show that $e = e'$.

2. **Uniqueness of inverse elements.** Let G be a group with identity element e. Let $g \in G$ and suppose that $g * h = h * g = g * h' = h' * g = e$. Show that $h = h'$.

3. **The cancellation law.** Suppose that $gx = hx$ for some elements g, h, x in a group G. Show that $g = h$. [Note that the same proof, mutatis mutandis, shows that if $xg = xh$, then $g = h$.]

4. **The "socks and shoes" property.** Let g, h be elements of a group G. Show that $(gh)^{-1} = h^{-1}g^{-1}$.

5. **Product Groups.** Given two groups G, H with group operations $*_G, *_H$, the Cartesian product $G \times H$ is a group with the operation $*_{G \times H}$ given by

$$(g, h) *_{G \times H} (g', h') = (g *_G g', h *_H h').$$

   Show that this operation satisfies the definition of a group.

6. **Cyclic groups.** A group G is called **cyclic** if there exists an element g in G, called a **generator**, such that the sequence

$$\left(g^k\right)_{k \in \mathbb{Z}} = (\ldots, g^{-3}, g^{-2}, g^{-1}, g^0, g^1, g^2, g^3, \ldots)$$

   contains all of the elements in G.

   (a) Suppose that a group G is finite and cyclic, with generator g. Show that

$$G = \{g, g^2, g^3, \ldots, g^{|G|}\}.$$

   (b) The group of integers is cyclic. Find all of the generators.

   (c) The group $\mathbb{Z}_8$ is cyclic. Find all of the generators.

   (d) The group $\mathbb{Z}_2 \times \mathbb{Z}_3$ is cyclic. Find all of the generators.

   (e) Show that the group $\mathbb{Z}_2 \times \mathbb{Z}_2$ is *not* cyclic.

   Hint. For part (a), let n be the least positive integer such that $g^n = e$ (explain why n exists!). Given an arbitrary element $h \in G$, write $h = g^k$ for some k, then use the Division Algorithm.

7. **Cyclic permutations.** Let k, n be positive integers with $k \leq n$, and let $A = \{a_1, a_2, \ldots, a_k\}$ be a set of k distinct elements in $\{1, 2, \ldots, n\}$. We write $(a_1 a_2 \cdots a_k)$ to denote the permutation σ in S_n (see Definition 2.1.1, p. 22) given by the assignments

$$a_1 \to a_2 \to a_3 \to \cdots \to a_k \to a_1$$

and $\sigma(j) = j$ for $j \notin A$. If $k = 1$, the permutation (a_1) is the identity permutation. A permutation of the form $(a_1 a_2 \cdots a_k)$ is called a k-**cycle**. For example, the element $\sigma = [1, 4, 2, 3, 5] = (2\ 4\ 3)$ is a 3-cycle in S_5 because σ acts on the set $A = \{2, 3, 4\}$ by

$$2 \to 4 \to 3 \to 2$$

and σ acts on $\{1, 5\}$ as the identity. Note that cycle notation is not unique. For example, we have $(2\ 4\ 3) = (4\ 3\ 2) = (3\ 2\ 4)$. A permutation is called **cyclic** if it is a k-cycle for some k. A 2-cycle is called a **transposition**.

   (a) Find all of the cyclic permutations in S_3. Find their inverses.

   (b) Find all of the cyclic permutations in S_4.

8. Cycles $(a_1 a_2 \cdots a_k)$ and $(b_1 b_2 \cdots b_\ell)$ are called **disjoint** if the sets $\{a_1, a_2, \ldots, a_k\}$ and $\{b_1, b_2, \ldots, b_\ell\}$ are disjoint, that is, if $a_i \neq b_j$ for all i, j. Show that every permutation in S_n is either a cycle or a product of disjoint cycles.

9. **Every permutation is a product of transpositions.**

   (a) Let $n \geq 2$. Show that every permutation in S_n can be written as a product of transpositions.

   (b) Show that factoring a permutation into a product of transpositions is not unique by writing the identity permutation in S_3 as a product of transpositions in two different ways.

10. **Cayley tables.** The **Cayley table** for a finite group G is a two-dimensional array with rows and columns labeled by the elements of the group, and with entry gh in position with row label g and column label h. Partial Cayley tables for S_3 (Figure 2.2.7, p. 31) and D_4 (Figure 2.2.8, p. 31) are given below.

	e	(23)	(13)	(12)	(123)	(132)
e				(12)		
(23)						
(13)		(132)				
(12)					(23)	
(123)						
(132)						

Figure 2.2.7 (Partial) Cayley table for S_3. The symbol e denotes the identity permutation.

	F_V	F_H	F_D	$F_{D'}$	$R_{1/4}$	$R_{1/2}$	$R_{3/4}$	R_0
F_V		$R_{1/2}$						
F_H					F_D			
F_D						$F_{D'}$		
$F_{D'}$								
$R_{1/4}$								
$R_{1/2}$								
$R_{3/4}$								
R_0								

Figure 2.2.8 (Partial) Cayley table for D_4. (See Checkpoint 2.1.6, p. 24 for notation for the elements of D_4.)

(a) Fill in the remaining entries in the Cayley tables for S_3 and D_4.

(b) Prove that the Cayley table for any group is a **Latin square**. This means that every element of the group appears exactly once in each row and in each column.

(See *Solutions to Exercises* in Appendix B, p. 103.)

2.3 Subgroups and cosets

Definition 2.3.1 Subgroups and cosets. A subset H of a group G is called a **subgroup** of G if H itself is a group under the group operation of G restricted to H. We write $H \leq G$ to indicate that H is a subgroup of G. Given a subgroup H of G, and given an element g in G, the set

$$gH := \{gh \colon h \in H\}$$

is called a **(left) coset** of H. The set of all cosets of H is denoted G/H, that is,

$$G/H := \{gH \colon g \in G\}.$$

$\Diamond$

Checkpoint 2.3.2 Consider D_4 as described in Checkpoint 2.1.6, p. 24.

$$D_4 = \{R_0, R_{1/4}, R_{1/2}, R_{3/4}, F_H, F_V, F_D, F_{D'}\}$$

1. Is the subset $\{R_0, R_{1/4}, R_{1/2}, R_{3/4}\}$ of rotations a subgroup of D_4? Why or why not?

2. Is the subset $\{F_H, F_V, F_D, F_{D'}\}$ of reflections a subgroup of D_4? Why or why not?

(See *Solutions to Exercises* in Appendix B, p. 103.)

Checkpoint 2.3.3 Find G/H for $G = S_3$, $H = \{e, (12)\}$.
(See *Solutions to Exercises* in Appendix B, p. 103.)

Proposition 2.3.4 Subgroup tests. *Let H be a subset of a group G. The following are equivalent.*

i H is a subgroup of G

ii (2-step subgroup test) H is nonempty, ab is in H for every a, b in H, and a^{-1} is in H for every a in H

iii (1-step subgroup test) H is nonempty and ab^{-1} is in H for every a, b in H

Proposition 2.3.5 Subgroup generated by a set of elements. *Let S be a nonempty subset of a group G, and let S^{-1} denote the set $S^{-1} = \{s^{-1} \colon s \in S\}$ of inverses of elements in S. We write $\langle S \rangle$ to denote the set of all elements of G of the form*

$$s_1 s_2 \cdots s_k$$

*where k ranges over all positive integers and each s_i is in $S \cup S^{-1}$ for $1 \leq i \leq k$. The set $\langle S \rangle$ is a subgroup of G, called the **subgroup generated by the set S** , and the elements of S are called the **generators** of $\langle S \rangle$.*

Comment on notational convention: If $S = \{s_1, s_2, \ldots, s_k\}$ is finite, we write $\langle s_1, s_2, \ldots, s_k \rangle$ for $\langle S \rangle$, instead of the more cumbersome $\langle \{s_1, s_2, \ldots, s_k\} \rangle$.

Observation 2.3.6 If G is a cyclic group with generator g, then $G = \langle g \rangle$.

Checkpoint 2.3.7 Show that $\langle S \rangle$ is indeed a subgroup of G. How would this fail if S were empty?

Checkpoint 2.3.8

1. Find $\langle F_H, F_V \rangle \subseteq D_4$.

2. Find $\langle 6, 8 \rangle \subseteq \mathbb{Z}$.

   (See *Solutions to Exercises* in Appendix B, p. 103.)

Proposition 2.3.9 Cosets as equivalence classes. *Let G be a group and let H be a subgroup of G. Let $\sim_H$ be the relation on G defined by $x \sim_H y$ if and only if $x^{-1}y \in H$. The relation $\sim_H$ is an equivalence relation on G, and the equivalence classes are the cosets of H, that is, we have $G/\sim_H = G/H$.*

Corollary 2.3.10 Cosets as a partition. *Let G be a group and let H be a subgroup of G. The set G/H of cosets of H form a partition of G.*

Exercises

1. Prove Proposition 2.3.4, p. 32. Be aware of the following subtlety: a subgroup H of a group G must have an identity element, but Definition 2.3.1, p. 32 does *not* require that it be the *same* as the identity element for G. Similarly, an element of H must have in inverse, but Definition 2.3.1, p. 32 does not require that it be the same as the inverse in G. Suggestion: start by proving a lemma that says if H is a subgroup of G, then the identity element for H must be the same as the identity element for G, and that inverses of elements in H are the same as inverses in G. Then use the lemma in your proof that statements (i), (ii), and (iii) are equivalent.

2. Find all the subgroups of S_3.
    (See *Solutions to Exercises* in Appendix B, p. 103.)

3. Find all the cosets of the subgroup $\{R_0, R_{1/2}\}$ of D_4.

4. **Subgroups of $\mathbb{Z}$ and $\mathbb{Z}_n$.**

    (a) Let H be a subgroup of $\mathbb{Z}$. Show that either $H = \{0\}$ or $H = \langle d \rangle$, where d is the smallest positive element in H.

    (b) Let H be a subgroup of $\mathbb{Z}_n$. Show that either $H = \{0\}$ or $H = \langle d \rangle$, where d is the smallest positive element in H.

(c) Let $n_1, n_2, \ldots, n_r$ be positive integers. Show that

$$\langle n_1, n_2, \ldots, n_r \rangle = \langle \gcd(n_1, n_2, \ldots, n_r) \rangle.$$

Consequence of this exercise: The greatest common divisor $\gcd(a, b)$ of integers a, b is the smallest positive integer of the form $sa + tb$ over all integers s, t. Two integers a, b are relatively prime if and only if there exist integers s, t such that $sa + tb = 1$.

Hint. Suggestion for part (c): Do the case $r = 2$ first.

5. **Centralizers, Center of a group.** The **centralizer** of an element a in a group G, denoted $C(a)$, is the set

$$C(a) = \{g \in G : ag = ga\}.$$

The **center** of a group G, denoted $Z(G)$, is the set

$$Z(G) = \{g \in G : ag = ga \ \forall a \in G\}.$$

(a) Show that the centralizer $C(a)$ of any element a in a group G is a subgroup of G.

(b) Show that the center $Z(G)$ of a group G is a subgroup of G.

6. **The order of a group element.** Let g be an element of a group G. The **order** of g, denoted $|g|$, is the smallest positive integer n such that $g^n = e$, if such an integer exists. If there is no positive integer n such that $g^n = e$, then g is said to have *infinite* order. Show that, if the order of g is finite, say $|g| = n$, then

$$\langle g \rangle = \{g^0, g^1, g^2, \ldots, g^{n-1}\}.$$

Consequence of this exercise: If G is cyclic with generator g, then $|G| = |g|$.

7. **Cosets of a subgroup partition the group, Lagrange's Theorem.**

(a) Prove Proposition 2.3.9, p. 33.

(b) Now suppose that a group G is finite. Show that all of the cosets of a subgroup H have the same size.

(c) Prove Lagrange's Theorem, stated below.

Lagrange's Theorem.

If G is a finite group and H is a subgroup, then the order of H divides the order of G.

Hint. For part (b), let aH, bH be cosets. Show that the function $aH \to bH$ given by $x \to ba^{-1}x$ is a bijection.

8.   **Consequences of Lagrange's Theorem.**

   (a) Show that the order of any element of a finite group divides the order of the group.

   (b) Let G be a finite group, and let $g \in G$. Show that $g^{|G|} = e$.

   (c) Show that a group of prime order is cyclic.

2.4 Group homomorphisms

Definition 2.4.1 Group homomorphism. Let G, H be groups, with group operations $*_G, *_H$, respectively. A map $\phi\colon G \to H$ is called a **homomorphism** if

$$\phi(x *_G y) = \phi(x) *_H \phi(y)$$

for all x, y in G. A homomorphism that is both injective (one-to-one) and surjective (onto) is called an **isomorphism** of groups. If $\phi\colon G \to H$ is an isomorphism, we say that G is **isomorphic** to H, and we write $G \approx H$. ◇

Checkpoint 2.4.2 Show that each of the following are homomorphisms.

- $GL(n, \mathbb{R}) \to \mathbb{R}^*$ given by $M \to \det M$

- $\mathbb{Z} \to \mathbb{Z}$ given by $x \to mx$, some fixed $m \in \mathbb{Z}$

- $G \to G$, G any group, given by $x \to axa^{-1}$, some fixed $a \in G$

- $\mathbb{C}^* \to \mathbb{C}^*$ given by $z \to z^2$

Show that each of the following are not homomorphisms. In each case, demonstrate what fails.

- $\mathbb{Z} \to \mathbb{Z}$ given by $x \to x + 3$

- $\mathbb{Z} \to \mathbb{Z}$ given by $x \to x^2$

- $D_4 \to D_4$ given by $g \to g^2$

Definition 2.4.3 Kernel of a group homomorphism. Let $\phi\colon G \to H$ be a group homomorphism, and let e_H be the identity element for H. We write $\ker(\phi)$ to denote the set

$$\ker(\phi) := \phi^{-1}(e_H) = \{g \in G \colon \phi(g) = e_H\},$$

called the **kernel** of ϕ. ◇

Checkpoint 2.4.4 Find the kernel of each of the following homomorphisms.

- $\mathbb{C}^* \to \mathbb{C}^*$ given by $z \to z^n$

- $\mathbb{Z}_8 \to \mathbb{Z}_8$ given by $x \to 6x \pmod 8$

- $G \to G$, G any group, given by $x \to axa^{-1}$, some fixed $a \in G$

(See *Solutions to Exercises* in Appendix B, p. 103.)

Proposition 2.4.5 Basic properties of homomorphisms. *Let $\phi\colon G \to H$ be a homomorphism of groups. Let e_G, e_H denote the identity elements of G, H, respectively. We have the following.*

1. $\phi(e_G) = e_H$

2. $\phi\left(g^{-1}\right) = (\phi(g))^{-1}$ *for all* $g \in G$

3. $\ker(\phi)$ *is a subgroup of* G

4. $\phi(G)$ *is a subgroup of* H

5. $\phi(x) = y$ *if and only if* $\phi^{-1}(y) = x\ker(\phi)$

6. $\phi(a) = \phi(b)$ *if and only if* $a\ker(\phi) = b\ker(\phi)$

7. ϕ *is one-to-one if and only if* $\ker(\phi) = \{e_G\}$

Proposition 2.4.6 G/K **is a group if and only if** K **is a kernel.** *Let K be a subgroup of a group G. The set G/K of cosets of K forms a group, called a* **quotient group** *(or* **factor group**)*, under the operation*

$$(xK)(yK) = xyK \tag{2.4.1}$$

if and only if K is the kernel of a homomorphism $G \to G'$ for some group G'.

Corollary 2.4.7 (First Isomorphism Theorem). *Let $\phi\colon G \to H$ be a homomorphism of groups. Then $G/\ker(\phi)$ is isomorphic to $\phi(G)$ via the map $g\ker(\phi) \to \phi(g)$.*

Definition 2.4.8 Normal subgroup. A subgroup H of a group G is called **normal** if $ghg^{-1} \in H$ for every $g \in G$, $h \in H$. We write $H \trianglelefteq G$ to indicate that H is a normal subgroup of G. ◇

Proposition 2.4.9 Characterization of normal subgroups. *Let K be a subgroup of a group G. The following are equivalent.*

1. K *is the kernel of some group homomorphism* $\phi\colon G \to H$

2. G/K *is a group with multiplication given by Equation (2.4.1)*

3. K *is a normal subgroup of* G

Exercises

Basic properties of homomorphisms. Prove Proposition 2.4.5, p. 36.

1. Prove Properties 1 and 2.

2. Prove Properties 3 and 4.

3. Prove Properties 5, 6, and 7.

   Hint. Use Fact 1.4.5, p. 14.

4. Show that the inverse of an isomorphism is an isomorphism.

Proof of the First Isomorphism Theorem.

5.   Prove Proposition 2.4.6, p. 37.

   Hint. First, suppose $K = \ker(\phi)$ for some homomorphism $\phi\colon G \to G'$. Explain why Item 6, p. 37 of Proposition 2.4.5, p. 36 can be rephrased to say that there is a one-to-one correspondence $G/K \leftrightarrow \phi(G)$ given by $gK \leftrightarrow \phi(g)$. Now use the bijection $G/K \leftrightarrow \phi(G)$ to impose the group structure of $\phi(G)$ (Item 4, p. 37 of Proposition 2.4.5, p. 36) on G/K. Conversely, if G/K is a group with the group operation (2.4.1), define $\phi\colon G \to G/K$ by $\phi(g) = gK$, then check that ϕ is a homomorphism and that $\ker(\phi) = K$.

6.   Prove Corollary 2.4.7, p. 37 by explicitly showing how it follows from the proof of Proposition 2.4.6, p. 37 outlined in Exercise 2.4.5, p. 38.

7.   Let n, a be relatively prime positive integers. Show that the map $\mathbb{Z}_n \to \mathbb{Z}_n$ given by $x \to ax$ is an isomorphism.

   Hint. Use the fact that $\gcd(m, n)$ is the least positive integer of the form $sm + tn$ over all integers s, t (see Exercise 2.3.4, p. 33). Use this to solve $ax = 1 \pmod{n}$ when a, n are relatively prime.

8.   **Another construction of $\mathbb{Z}_n$.** Let $n \geq 1$ be an integer and let $\omega = e^{i2\pi/n}$. Let $\phi\colon \mathbb{Z} \to S^1$ be given by $k \to \omega^k$.

   (a) Show that the the image of ϕ is the group C_n of nth roots of unity.

   (b) Show that ϕ is a homomorphism, and that the kernel of ϕ is the set $n\mathbb{Z} = \{nk\colon k \in \mathbb{Z}\}$.

   (c) Conclude that $\mathbb{Z}/(n\mathbb{Z})$ is isomorphic to the group of n-th roots of unity.

9.   **Isomorphic images of generators are generators.** Let S be a subset of a group G. Let $\phi\colon G \to H$ be an isomorphism of groups, and let $\phi(S) = \{\phi(s)\colon s \in S\}$. Show that $\phi(\langle S \rangle) = \langle \phi(S) \rangle$.

10. **Conjugation.** Let G be a group, let a be an element of G, and let $C_a\colon G \to G$ be given by $C_a(g) = aga^{-1}$. The map C_a is called **conjugation** by the element a and the elements g, aga^{-1} are said to be **conjugate** to one another.

   (a) Show that C_a is an isomorphism of G with itself.

   (b) Show that "is conjugate to" is an equivalence relation. That is, consider the relation on G given by $x \sim y$ if $y = C_a(x)$ for some a. Show that this is an equivalence relation.

11. **Isomorphism induces an equivalence relation.** Prove that "is isomorphic to" is an equivalence relation on groups. That is, consider the relation $\approx$ on the set of all groups, given by $G \approx H$ if there exists a group isomorphism $\phi\colon G \to H$. Show that this is an equivalence relation.

Characterization of normal subgroups. Prove Proposition 2.4.9, p. 37. (Note that the equivalence of Item 1, p. 37 and Item 2, p. 37 has already been established by Proposition 2.4.6, p. 37.)

12. Show that Item 1, p. 37 implies Item 3, p. 37.

13. Show that Item 3, p. 37 implies Item 2, p. 37. The messy part of this proof is to show that multiplication of cosets is well-defined. This means you start by supposing that $xK = x'K$ and $yK = y'K$, then show that $xyK = x'y'K$.

14. **Further characterizations of normal subgroups.** Show that Item 3, p. 37 is equivalent to the following conditions.

    (a) $gKg^{-1} = K$ for all $g \in G$

    (b) $gK = Kg$ for all $g \in G$

15. **Automorphisms.** Let G be a group. An **automorphism** of G is an isomorphism from G to itself. The set of all automorphisms of G is denoted $\text{Aut}(G)$.

    (a) Show that $\text{Aut}(G)$ is a group under the operation of function composition.

    (b) Show that
    $$\text{Inn}(G) := \{C_g \colon g \in G\}$$
    is a subgroup of $\text{Aut}(G)$. (The group $\text{Inn}(G)$ is called the group of **inner automorphisms** of G.)

    (c) Find an example of an automorphism of a group that is not an inner automorphism.

2.5 Group actions

Definition 2.5.1 Group action, orbit, stabilizer. Let G be a group and let X be a set. An **action** of the group G on the set X is a group homomorphism

$$\phi \colon G \to \mathrm{Perm}(X).$$

We say that the group G **acts** on the set X, and we call X a G-**space**. For $g \in G$ and $x \in X$, we write gx to denote $(\phi(g))(x)$.[1] We write $\mathrm{Orb}(x)$ to denote the set

$$\mathrm{Orb}(x) = \{gx \colon g \in G\},$$

called the **orbit** of x, and we write $\mathrm{Stab}(x)$ to denote the set

$$\mathrm{Stab}(x) = \{g \in G \colon gx = x\},$$

called the **stabilizer** or **isotropy** subgroup[2] of x. A group action is **transitive** if there is only one orbit. A group action is **faithful** if the map $G \to \mathrm{Perm}(X)$ has a trivial kernel. ◇

Checkpoint 2.5.2 Find the indicated orbits and stabilizers for each of the following group actions.

1. D_4 acts on the square $X = \{(x, y) \in \mathbb{R}^2 \colon -1 \le x, y \le 1\}$ by rotations and reflections. What is the orbit of $(1, 1)$? What is the orbit of $(1, 0)$? What is the stabilizer of $(1, 1)$? What is the stabilizer of $(1, 0)$?

2. The circle group S^1 (see Subsection 2.1.3, p. 25) acts on the two-sphere S^2 by rotation about the z-axis. Given an element $e^{i\alpha}$ in S^1 a point (θ, ϕ) in S^2 (in spherical coordinates), the action is given by

   $$e^{i\alpha} \cdot (\theta, \phi) = (\theta, \phi + \alpha).$$

   What is the orbit of $(\pi/4, \pi/6)$? What is the orbit of the north pole $(0, 0)$? What is the stabilizer of $(\pi/4, \pi/6)$? What is the stabilizer of the north pole?

3. Any group G acts on itself by *conjugation*, that is, by $(\phi(g))(x) = gxg^{-1} = C_g(x)$ (see Exercise 2.4.10, p. 38). Describe the orbit and stabilizer of a group element x.

(See *Solutions to Exercises* in Appendix B, p. 103.)

Checkpoint 2.5.3 Show that the stabilizer of an element x in a G-space X is a subgroup of G.

---

[1] Other notations for $(\phi(g))(x)$ are $g(x)$, $g \cdot x$, and $g.x$.
[2] It must be proved that $\mathrm{Stab}(x)$ is indeed a subgroup of G. See Checkpoint 2.5.3, p. 40 below.

Definition 2.5.4 Orbit space. We write X/G to denote the set

$$X/G = \{\mathrm{Orb}(x) : x \in X\}$$

of orbits of the group G acting on a set X. The set X/G is also called the **orbit space** of the group action. ◇

Checkpoint 2.5.5 Describe X/G for each of the three group actions in Checkpoint 2.5.2, p. 40.

Proposition 2.5.6 *Let group G act on set X. Let $\sim_G$ denote the relation on X given by $x \sim_G y$ if and only if $y = gx$ for some $g \in G$. Then we have that $\sim_G$ is an equivalence relation, and further, we have $x \sim_G y$ if and only if $y \in \mathrm{Orb}(x)$. In other words, we have*

$$X/G = X/\sim_G .$$

In particular, we have that the orbit space X/G is a partition of X.

Theorem 2.5.7 The Orbit-Stabilizer Theorem. *Let G be a group acting on a set X, and let x be an element of X. There is a one-to-one correspondence*

$$G/\mathrm{Stab}(x) \leftrightarrow \mathrm{Orb}(x)$$

given by

$$g\,\mathrm{Stab}(x) \leftrightarrow gx.$$

Exercises

1. **The sign of a permutation.** Let n be a positive integer, and let Δ_n be the polynomial

$$\Delta_n = \prod_{1 \le i < j \le n} (x_i - x_j)$$

in the variables $x_1, x_2, \ldots, x_n$. For example, we have $\Delta_3 = (x_1 - x_2)(x_1 - x_3)(x_2 - x_3)$. Given a permutation α in S_n, let $\alpha\Delta_n$ be the polynomial

$$\alpha\Delta_n = \prod_{1 \le i < j \le n} (x_{\alpha(i)} - x_{\alpha(j)}).$$

In the exercises below, you will show that $\alpha\Delta_n = \pm\Delta_n$ for all α in S_n. This allows us to define the **sign** of a permutation α, denoted $\mathrm{sgn}(\alpha)$, to be $+1$ or -1 according to whether $\alpha\Delta_n = \Delta_n$ or $\alpha\Delta_n = -\Delta_n$, respectively.

$$\mathrm{sgn}(\alpha) = \begin{cases} +1 & \text{if } \alpha\Delta_n = \Delta_n \\ -1 & \text{if } \alpha\Delta_n = -\Delta_n \end{cases}$$

We say that a permutation α is **even** if $\mathrm{sgn}(\alpha) = +1$, and we say α is **odd** if $\mathrm{sgn}(\alpha) = -1$.

(a) Write expressions for Δ_4, $(134)\Delta_4$, and $(1324)\Delta_4$ (write all the factors). All three expressions consist of the same factors, possibly times ± 1. Identify which factors of $(134)\Delta_4$ and $(1324)\Delta_4$ have the opposite sign of the corresponding factor in Δ_4. Use your expressions to find $\text{sgn}(134)$ and $\text{sgn}(1324)$.

(b) Generalize the examples in part (a) above to show that $\alpha\Delta_n = \pm\Delta_n$ for all α in S_n. This justifies the definition of the sign of a permutation.

(c) Show that $\tau\Delta_n = -\Delta_n$ for any transposition τ in S_n. Suggestion: Let $\tau = (ab)$ be a transposition with $a < b$. Count the number of factors $(x_i - x_j)$ of Δ_n such that $x_{\tau(i)} - x_{\tau(j)} = -(x_i - x_j)$.

(d) Define $\alpha(-\Delta_n)$ by $\alpha(-\Delta_n) = -\alpha\Delta_n$ for α in S_n. Show that

$$(\alpha\beta)\Delta_n = \alpha(\beta\Delta_n) \qquad (2.5.1)$$

for all α, β in S_n.

(e) Let α be an element of S_n. By Exercise 2.2.9, p. 30, α may be written as a product of transpositions $\alpha = \tau_1\tau_2\cdots\tau_k$. Use $(2.5.1)$ to show that $\text{sgn}(\alpha) = (-1)^k$. Consequence: if α is expressible as a product of an even number of transpositions, then α is an even permutation. Further, any product of transpositions that equals α must have an even number of transpositions. (A similar statement holds replacing the word "even" by the word "odd".)

(f) Show that $\phi(\alpha)(\pm\Delta_n) = \pm\alpha\Delta_n$ defines a group action $\phi\colon S_n \to \text{Perm}(X)$ of the group S_n on the set $X = \{\Delta_n, -\Delta_n\}$.

(g) Show that the sign function $\text{sgn}\colon S_n \to C_2$ is a homomorphism of groups.

(h) The subset of even permutations of S_n is denoted A_n. Give two arguments why A_n is a normal subgroup of S_n. Use (i) the 1-step or the 2-step subgroup test and definition Definition 2.4.8, p. 37, and (ii) using criterion 1 of Proposition 2.4.9, p. 37. This group A_n is called the **alternating group**.

Hint. For part (d): Both sides of $(2.5.1)$ are equal to $\prod_{i<j}\left(x_{\alpha\beta(i)} - x_{\alpha\beta(j)}\right)$.

2. **Actions of a group on itself.** Let G be a group. Here are three actions $G \to \text{Perm}(G)$ of G on itself. **Left multiplication** L is given by

$$g \to L_g$$

where L_g is given by $L_g(h) = gh$. **Right inverse multiplication** R

is given by
$$g \to R_g$$
where R_g is given by $R_g(h) = hg^{-1}$. **Conjugation** C is given by
$$g \to C_g$$
where C_g is given by $C_g(h) = ghg^{-1}$.

(a) Show that, for $g \in G$, the maps L_g, R_g, C_g are elements of Perm(G).

(b) Show that each of these maps L, R, C is indeed a group action.

(c) Show that the map L is injective, so that $G \approx L(G)$.

Consequence of this exercise: Every group is isomorphic to a subgroup of a permutation group.

3.   **Cosets, revisited.** Let H be a subgroup of a group G, and consider the map
$$R \colon H \to \mathrm{Perm}(G)$$
given by $h \to R_h$, where $R_h(g) = gh^{-1}$ (this is the restriction of right inverse multiplication action in Exercise 2.5.2, p. 42 to H). Show that the orbits of this action of H on G are the same as the cosets of H. This shows that the two potentially different meanings of G/H (one is the set of cosets, the other is the set of orbits of the action of H on G via R), are in fact in agreement.

4.   **The natural action of a matrix group on a vector space.** Let G be a group whose elements are $n \times n$ matrices with entries in a field $\mathbb{F}$ and with the group operation of matrix multiplication. The **natural action** $G \to \mathrm{Perm}(X)$ of G on the vector space $X = \mathbb{F}^n$ is given by
$$g \to [v \to g \cdot v],$$
where the "dot" in the expression $g \cdot v$ is ordinary multiplication of a matrix times a column vector. Show that this is indeed a group action.

5.   Prove Proposition 2.5.6, p. 41.

(a) Show that $\sim_G$ is an equivalence relation.

(b) Show that $x \sim_G y$ if and only if $y \in \mathrm{Orb}(x)$.

(c) Explain why X/G is a partition of X.

6.   Prove The Orbit-Stabilizer Theorem (Theorem 2.5.7, p. 41).
Hint.   Let $x \in X$, and define a map $f_x \colon G \to X$ given by $f_x(g) = gx$.

Apply Fact 1.4.5, p. 14 to get a one-to-one correspondence

$$G/\sim_{f_x} \leftrightarrow f_x(G) = \mathrm{Orb}(x).$$

Then use Proposition 2.3.9, p. 33 to show that $G/\sim_{f_x} = G/\mathrm{Stab}(x)$. Whatever proof method you use, you must address the issue of well-definedness of one of the maps $g\,\mathrm{Stab}(x) \to gx$ or $gx \to g\,\mathrm{Stab}(x)$.

7. **The projective linear group action on projective space.** Let V be a vector space over a field $\mathbb{F}$ (in this course, the base field $\mathbb{F}$ is either the real numbers $\mathbb{R}$ or the complex numbers $\mathbb{C}$). The group $\mathbb{F}^*$ of nonzero elements in $\mathbb{F}$ acts on the set $V \setminus \{0\}$ of nonzero elements in V by scalar multiplication, that is, by the map $\alpha \to [v \to \alpha v]$. The set of orbits $(V \setminus \{0\})/\mathbb{F}^*$ is called the *projectivization* of V, or simply **projective space**, and is denoted $\mathbb{P}(V)$.

(a) Let $\sim_{\mathrm{proj}}$ denote the equivalence relation that defines the orbits $(V \setminus \{0\})/\mathbb{F}^*$. Verify that $\sim_{\mathrm{proj}}$ is given by $x \sim_{\mathrm{proj}} y$ if and only if $x = \alpha y$ for some $\alpha \in \mathbb{F}^*$.

(b) Verify that the group $GL(V)$ (the group of invertible linear transformations of V) acts on $\mathbb{P}(V)$ by

$$g \cdot [v] = [g(v)] \qquad\qquad (2.5.2)$$

for $g \in GL(V)$ and $v \in V \setminus \{0\}$.

(c) Show that the kernel of the map $GL(V) \to \mathrm{Perm}(\mathbb{P}(V))$ given by (2.5.2) is the subgroup $K = \{\alpha\,\mathrm{Id} : \alpha \in \mathbb{F}^*\}$.

(d) Conclude that the **projective linear group** $PGL(V) := GL(V)/K$ acts on $\mathbb{P}(V)$.

(e) Show that $\mathbb{F}^*$ acts on $GL(V)$ by $\alpha \cdot T = \alpha T$, and that $PGL(V) \approx GL(V)/\mathbb{F}^*$.

(f) Let $s\colon S^2 \to \hat{\mathbb{C}}$ denote the stereographic projection (see (1.3.4)). Show that the map $\mathbb{P}(\mathbb{C}^2) \to S^2$ given by $[(\alpha, \beta)] \to s^{-1}(\alpha/\beta)$ if $\beta \neq 0$ and given by $[(\alpha, \beta)] \to (0, 0, 1)$ if $\beta = 0$ is well-defined and is a bijection.

2.6 Additional exercises

Exercises

1. **The group of units in $\mathbb{Z}_n$.** Let U_n denote the set of elements in $\mathbb{Z}_n$ that have multiplicative inverses, that is,

    $$U_n = \{x \in \mathbb{Z}_n : \exists y, xy = 1 \pmod{n}\}.$$

    (a) Show that x is in U_n if and only if x is relatively prime to n.

    (b) Show that U_n with the binary operation of multiplication mod n is an Abelian group.

    (c) Show that U_n is isomorphic to $\mathrm{Aut}(\mathbb{Z}_n)$ via $x \to [a \to ax]$.

    Terminology: The group U_n is called the the group of (**multiplicative**) **units** in $\mathbb{Z}_n$. The function $n \to |U_n|$, important in number theory, is called the **Euler phi function**, written $\phi(n) = |U_n|$.

2. **Fermat's Little Theorem.** For every integer x and every prime p, we have $x^p = x \pmod{p}$.

    Hint. First, reduce x mod p, that is, write $x = qp + r$ with $0 \le r \le p - 1$. Now consider two cases. The case $r = 0$ is trivial. If $r \ne 0$, apply the fact $r^{|G|} = e$ (see Exercise 2.3.8, p. 35) to the group $G = U_p$.

3. **An alternative approach to parity of permutations.** This exercise gives a way to define even and odd permutations that is different from the method used in Exercise 2.5.1, p. 41.

    (a) Suppose that the identity permutation e in S_n is written as a product of transpositions

    $$e = \tau_1 \tau_2 \cdots \tau_r.$$

    Show that r is even.

    (b) Suppose that σ in S_n is written in two ways as a product of transpositions.

    $$\sigma = (a_1 b_1)(a_2 b_2) \cdots (a_s b_s) = (c_1 d_1)(c_2 d_2) \cdots (c_t d_t)$$

    Show that s, t are either both even or both odd. The common evenness or oddness of s, t is called the **parity** of the permutation σ.

    (c) Show that the parity of a k-cycle is even if k is odd, and the parity of a k-cycle is odd if k is even.

Hint.

(a) Consider the two rightmost transpositions $\tau_{r-1}\tau_r$. They have one of the following forms, where a, b, c, d are distinct.

$$(ab)(ab), (ac)(ab), (bc)(ab), (cd)(ab)$$

The first allows you to reduce the transposition count by two by cancelling. The remaining three can be rewritten.

$$(ab)(bc), (ac)(cb), (ab)(cd)$$

Notice that the index of the rightmost transposition in which the symbol a occurs has been reduced by 1 (from r to $r - 1$). Finish this reasoning with an inductive argument.

4. **The size of the alternating group.** Show that $|A_n| = |S_n|/2$ for $n \geq 2$, that is, that half of the elements of S_n are even, and half are odd.

5. **The order of a permutation.** Let $\sigma \in S_n$ be written as a product of disjoint cycles. Show that the order σ is the least common multiple of the lengths of those disjoint cycles.

6. **Alternative approach to multiplication in a factor group.** Given subsets S, T of a group G, define the set ST by

$$ST = \{st \colon s \in S, t \in T\}.$$

Now suppose that H is a subgroup of G. Show that $(xH)(yH) = xyH$ for all x, y in G if and only if H is a normal subgroup of G.

7. **Semidirect product.** Let K, H be groups, and let $\phi \colon H \to \mathrm{Aut}(K)$ be a homomorphism. The **semidirect product**, denoted $K \rtimes_\phi H$, or $K \rtimes H$ if ϕ is understood, is the set consisting of all pairs (k, h) with $k \in K$, $h \in H^1$ with the group multiplication operation $*$ given by

$$(k_1, h_1) * (k_1, h_2) = (k_1\phi(h_1)(k_2), h_1 h_2).$$

Two examples demonstrate why this is a useful construction. The dihedral group D_n is (isomorphic to) the semidirect product $C_n \rtimes C_2$, where C_n is the cyclic group generated by the rotation $R_{1/n}$ (rotation by $1/n$ of a revolution) and C_2 is the two-element group generated by any reflection R_L in D_n. The map $\phi \colon C_2 \to \mathrm{Aut}(C_n)$ is given by $F_L \to [R_\theta \to R_{-\theta}]$. The *Euclidean group* of congruence transformations of the plane is (isomorphic to) the group $\mathbb{R}^2 \rtimes O(2)$, where $(\mathbb{R}^2, +)$ is the additive group of 2×1 column vectors with real entries, and $O(2)$ is the group of 2×2 real orthogonal matrices. The map $\phi \colon O(2) \to \mathrm{Aut}(\mathbb{R}^2)$

is given by $g \to [v \to gv]$, that is to say, the natural action of $O(2)$ on $\mathbb{R}^2$. [The Euclidean group element (v, g) acts on the point $x \in \mathbb{R}^2$ by $x \to gx + v$.]

(a) Do all the necessary details to show that $K \rtimes H$ is indeed a group.

(b) *(Characterization of semidirect products)* Suppose that K, H are subgroups of a group G. Let $KH = \{kh \colon k \in K, h \in H\}$. Suppose that K is a normal subgroup of G, that $G = KH$, and that $K \cap H = \{e\}$. Show that $\phi \colon H \to \mathrm{Aut}(K)$, given by $\phi(h)(k) = hkh^{-1}$, is a homomorphism. Show that $\psi \colon K \times_\phi H \to G$, given by $\psi(k, h) = kh$, is an isomorphism.

(c) Show that $D_n \approx C_n \rtimes C_2$, as described above.

(d) Show that the following requirement holds for the Euclidean group action. We have

$$[(v_1, g_1)(v_2, g_2)]x = (v_1, g_1)[(v_2, g_2)x]$$

for all $v_1, v_2, x \in \mathbb{R}^2$ and $g_1, g_2 \in O(2)$.

(e) Suppose that $\phi \colon H \to \mathrm{Aut}(K)$ is the trivial homomorphism (that is, $\phi(h)$ is the identity homomorphism on K, for all $h \in H$). Show that $K \times_\phi H \approx K \times H$ in this case.

8. **Group action on functions on a G-space.** Suppose that a group G acts on a set X. Let $\mathcal{F}(X, Y)$ denote the set of functions

$$\mathcal{F}(X, Y) = \{f \colon X \to Y\}$$

from X to some set Y. Show that the formula

$$(g \cdot \alpha)(x) = \alpha(g^{-1} \cdot x)$$

defines an action of G on $\mathcal{F}(X, Y)$, where $g \in G$, $\alpha \in \mathcal{F}(X, Y)$, and $x \in X$.

---

[1]The notation $K \times H$ is understood to be the *direct* product group, so we do not use the Cartesian product notation to describe semidirect product, to avoid confusion, even though the underlying set for the direct product and the semidirect product are in fact the same Cartesian product $K \times H$.

Chapter 3

Geometries

3.1 Geometries and models

An integral part of the modern understanding of geometry is the concept of *congruence transformation*, or simply *symmetry*. The symmetries of a geometric space preserve inherent properties of figures, such as distance, angle, and area. In his 1872 work called the *Erlanger Programm*,[1] Felix Klein unified the study of a wide variety of geometric spaces by overtly placing the group of congruence transformations as part of the structure of a geometry. The following is a version of Klein's definition of geometry.

Definition 3.1.1 A **geometry** is a pair (X, G), where X is a set, called the **(model) space**, and G is a group, called the group of **congruence transformations**, that acts on X. Subsets of X are called **figures**. Figures F, F' are called **congruent** if there is an element g in G such that $g(F) = F'$. We write $F \cong F'$ to denote that figures F, F' are congruent.   ◇

Note on terminology and notation: Throughout this chapter on geometry, the term *transformation* will always mean a one-to-one and onto map of a space to itself. Given a geometry (X, G) with group action $\phi \colon G \to \mathrm{Perm}(X)$, we will abuse notation and write $g \colon X \to X$ to denote the map $\phi(g) \colon X \to X$ for an element g in G. It is common usage to say "the transformation g" to mean "the transformation $\phi(g)$" of the space X.

Checkpoint 3.1.2 Show that congruence is an equivalence relation on the set of figures in a geometry.

3.1.1 Examples of geometries

- **Planar Euclidean geometry**. The model space for planar Euclidean geometry is the plane $\mathbb{R}^2$. The group of congruence transformations consists of translations, rotations, reflections, and their compositions. Specifically, Euclidean congruences are functions of the form $v \to Rv + b$, where $v \in \mathbb{R}^2$, R is an element of the group of 2×2 orthogonal matrices, and $b \in \mathbb{R}^2$.

- **Spherical geometry**. The model space for spherical geometry is the sphere $S^2 = \{(x, y, z) \in \mathbb{R}^3 \colon x^2 + y^2 + z^2 = 1\}$. The group of congruence transformations consists of rotations of the sphere and reflections across planes through the origin. Specifically, spherical congruences are functions of the form $v \to Rv$, where $v \in \mathbb{R}^3$, $|v| = 1$, and R is an element of the group of 3×3 orthogonal matrices.

- **Projective geometry**. The model space for a projective geometry is projective space $\mathbb{P}(V)$, where V is a vector space V (see Exer-

---

[1]The German term *Erlanger Programm* (or "Erlangen program" in English) is named after the city Erlangen, where Klein worked at the university.

cise 2.5.7, p. 44 in the previous chapter). The group of congruence transformations is the projective linear group $PGL(V)$.

3.1.2 Planar geometries

Much of this chapter on geometry is devoted to a family of *planar* geometries whose model spaces are the extended complex plane $\hat{\mathbb{C}} = \mathbb{C} \cup \{\infty\}$ (described in section Section 1.3, p. 10) and some of its subsets. One of the properties shared by the congruence transformations in all of these planar geometries is *conformality*, or angle preservation. To say that a transformation T is **conformal** means that if two curves C_1 and C_2 make an oriented angle θ at a point P of intersection, then the two image curves $T(C_1)$ and $T(C_2)$ also make the same oriented angle at the point $T(P)$ of intersection (the angle made by two curves is the angle made by their tangents at the point of intersection). See Figure 3.1.3, p. 51. Exercise Group 3.1.4.2–5, p. 53 examines the conformal properties of the four basic types of complex functions summarized in Table 3.1.4, p. 51.

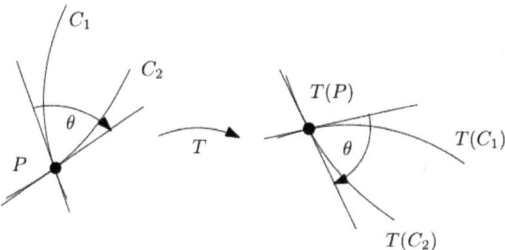

Figure 3.1.3 Conformal maps preserve oriented angles

Table 3.1.4 Basic types of transformations $\mathbb{C} \to \mathbb{C}$

homothety	$z \to kz, \ k > 0$
rotation	$z \to e^{it}z, \ t \in \mathbb{R}$
translation	$z \to z + b, \ b \in \mathbb{C}$
inversion	$z \to \frac{1}{z}$

Remark on convention: If $\alpha \in \mathbb{C}$ and if we write $\alpha > 0$, this implies that we mean that α is *real* and positive. This remark pertains to the definition of *homothety*[2] in Table 3.1.4, p. 51 above, and in the classification of normal forms, summarized in Table 3.2.28, p. 61 in the next section.

---

[2]Some texts define homothety to mean that k can be *any* nonzero real scalar. We follow the convention from complex analysis that restricts homotheties to positive real scalars. For example, see [1].

3.1.3 Subgeometries and equivalent geometries

Definition 3.1.5 Subgeometry. We say that a geometry (X, G) is a **subgeometry** of a geometry (Y, H) if X is a subset of Y and G is a subgroup of H and the action of G on X is compatible with the action of H on Y in the sense that $g \cdot x$ has the same value in both geometries (X, G) and (Y, H) for all $g \in G, x \in X$. $\Diamond$

Definition 3.1.6 Equivalent geometries. Geometries (X, G) and (Y, H) are **equivalent** if there is a bijective map $\mu \colon X \to Y$ such that every element of H has a conjugate transformation in G and every element of G has a conjugate transformation in H. In symbols:[3]

- for every $g \in G$, there is an $h \in H$ such that $\mu \circ g \circ \mu^{-1} = h$, and

- for every $h \in H$, there is a $g \in G$ such that $\mu^{-1} \circ h \circ \mu = g$.

Equivalent geometries are said to be *models* of the same geometry. $\Diamond$

  Note on terminology: The term "geometry" is used to refer to a specific model as in Definition 3.1.1, p. 50, and also to refer to an equivalence class of geometries.[4]

3.1.4 Exercises

1. **Warm up exercises with the three example geometries.**

    (a) Find the Euclidean congruence transformation that takes the triangle with vertices $(2, 0), (6, 0), (6, 3)$ to the triangle with vertices $(0, -1), (0, -5), (3, -1)$.

    (b) Find all possible spherical congruences that take the three points $(0, 0, 1), (0, 0, -1), (1, 0, 0)$ to the three points $(1, 0, 0), (-1, 0, 0), (0, 1, 0)$ (in that order). One of these is a rotation: find an axis and angle for that rotation.

    (c) Find the projective transformation in $PGL(2, \mathbb{C})$ that takes the three points $[(1, 1)], [(0, 1)], [(1, 0)]$ in $\mathbb{P}(\mathbb{C}^2)$ to $[(a, 1)], [(b, 1)], [(c, 1)]$ (in that order).

    (d) Find formulas for the composition of two Euclidean transformations and the inverse of a Euclidean transformation.

    (e) Let $d(P, Q)$ denote the distance between points P, Q in Euclidean

---

[3]See the note on terminology at the beginning of this section: the equation $\mu \circ g \circ \mu^{-1} = h$ below really means $\mu \circ \phi(g) \circ \mu^{-1} = \psi(h)$, where ϕ, ψ are the actions of G on X and H on Y, respectively.

[4]This is similar to the situation for groups, where the term "the cyclic group of order n" can refer to additive group $\mathbb{Z}_n$, or the multiplicative group of nth complex roots of unity, or simply the isomorphism class of these groups.

geometry, and let T be a Euclidean congruence transformation. Show that $d(T(P), T(Q)) = d(P, Q)$.

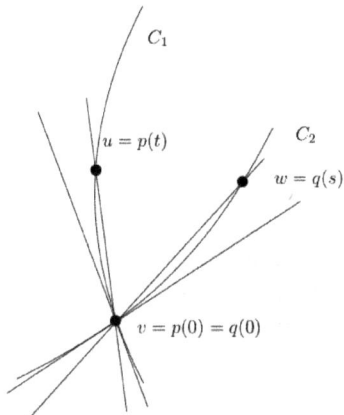

Figure 3.1.7 The angle between tangents to curves as a limit of secant approximations

Angles and conformal transformations.. The complex plane comes with a built-in measure of oriented angle. If u is a positive real number, $v = 0$, and $w \neq 0$ is a complex number, the measure of the oriented angle $\angle uvw$ is arg w. More generally, if u, v, w are three complex numbers with v distinct from u and w, the measure of the oriented angle $\angle uvw$ is

$$\arg\left(\frac{w - v}{u - v}\right). \tag{3.1.1}$$

2. Use the fact that rotations and translations are conformal to prove (3.1.1).

3. Use (3.1.1) to prove that homotheties are conformal.

4. Now suppose two curves C_1, C_2 intersect at v, let u be a point on C_1 and let w be a point on C_2. If u and w are close to v, then $\angle uvw$ is a secant approximation of an angle made by the tangents to C_1, C_2 at v. See Figure 3.1.7, p. 53. Now let $p(t), q(s)$ be parameterizations of C_1, C_2, respectively, with $p(0) = q(0) = v$, and $p(t_1) = u$, $q(s_1) = w$ for some $t_1, s_1 > 0$. We can compute an angle made by the tangents to the curves by the following limit.

$$\lim_{s \to 0^+, t \to 0^+} \arg\left(\frac{q(s) - v}{p(t) - v}\right) \tag{3.1.2}$$

The value of limit (3.1.2) is sensitive to the direction of the curve parameterizations and the sided-ness of the limits $t \to 0^{\pm}$ or

$s \to 0^{\pm}$. If the value of the limit (3.1.2) is θ for one set of choices for parameterizations and sided-ness, the limit for the other choices will be θ or $\theta \pm \pi$. For a given pair of parameterizations p, q, draw a sketch to illustrate the four possible cases $t \to 0^{\pm}, s \to 0^{\pm}$.

5. Use (3.1.1) and (3.1.2) to prove that inversion is conformal.

Equivalent geometries.

6. Show that the definition of equivalence of geometries actually defines an equivalence relation on geometries.

7. Let $\mathcal{E}_1$ denote the set of Euclidean congruence transformations given above in Subsection 3.1.1, p. 50. Let $\mathcal{E}_2$ denote the set of complex functions that can be written as compositions of the following three types.

    - $z \to e^{it}z$ for some $t \in \mathbb{R}$

    - $z \to z + b$ for some $b \in \mathbb{C}$

    - $z \to z^{*}$

    Show that the geometries $(\mathbb{R}^2, \mathcal{E}_1)$ and $(\mathbb{C}, \mathcal{E}_2)$ are equivalent.

8. Suppose that (X, G) and (Y, H) are equivalent geometries. Is it necessarily the case that G and H are isomorphic groups? If yes, give a proof. If no, give a counterexample.

3.2 Möbius geometry

Möbius geometry provides a unifying framework for studying planar geometries. In particular, the transformation groups of hyperbolic and elliptic
geometries in the sections that follow are subgroups of the group of Möbius
transformations.

3.2.1 Möbius transformations

A **Möbius transformation** (also called a **linear fractional transformation**) is a function of the form

$$f(z) = \frac{az + b}{cz + d} \qquad (3.2.1)$$

where z is a complex variable, a, b, c, d are complex constants, and $ad - bc \neq 0$.

Checkpoint 3.2.1

1. Show that if $c = 0$, then (3.2.1) defines a one-to-one and onto map
   $f : \mathbb{C} \to \mathbb{C}$ by finding an explicit inverse g for f and showing that $f \circ g$
   and $g \circ f$ are both the identity function on $\mathbb{C}$.

2. Show that if $c \neq 0$, then (3.2.1) defines a one-to-one and onto map
   $f : \mathbb{C} \setminus \{-d/c\} \to \mathbb{C} \setminus \{a/c\}$ by finding an explicit inverse g for f and
   showing that $f \circ g$ is the identity function on $\mathbb{C} \setminus \{a/c\}$ and $g \circ f$ is
   the identity function on $\mathbb{C} \setminus \{-d/c\}$.

3. For both cases 1 and 2 above, identify precisely where you use the
   condition $ad - bc \neq 0$. Show that, in fact, f is invertible if and only if
   $ad - bc \neq 0$.

   A Möbius transformation $f(z) = \frac{az+b}{cz+d}$ determines a one-to-one and
onto map of the extended complex plane $\hat{\mathbb{C}}$ to itself with the following
assignments: if $c = 0$, we define $f(\infty) = \infty$; if $c \neq 0$, we define $f(-d/c) = \infty$
and $f(\infty) = a/c$.

Checkpoint 3.2.2 Show that the composition of two Möbius transformations is a Möbius transformation. Suggestion: First show that the composition has the form $z \to \frac{rz+s}{tz+u}$. Next, instead of a brute force calculation to
check that $ru - ts \neq 0$, use Checkpoint 3.2.1, p. 55.

Definition 3.2.3 The definitions and Checkpoint exercises above show that
the set of all Möbius transformations forms a group under the operation of
composition of functions. This group, denoted MOB, is called the **Möbius
group**. **Möbius geometry** is the pair $(\hat{\mathbb{C}}, \text{MOB})$.   ◇

Note on notational convention: It is customary to use capital letters
such as S, T, U to denote Möbius transformations. It is also customary to

omit the parentheses, and to write Tz instead of $T(z)$ to denote the value of a Möbius transformation.

There is a natural relationship between Möbius group operations and matrix group operations. The map $\mathcal{T}\colon GL(2,\mathbb{C}) \to$ MOB given by

$$\begin{bmatrix} a & b \\ c & d \end{bmatrix} \to \left[z \to \frac{az+b}{cz+d} \right] \tag{3.2.2}$$

is a group homomorphism. The kernel of $\mathcal{T}$ is the group of nonzero scalar matrices, that is,

$$\ker \mathcal{T} = \left\{ \begin{bmatrix} k & 0 \\ 0 & k \end{bmatrix}, k \neq 0 \right\}.$$

The quotient group $GL(2,\mathbb{C})/\ker\mathcal{T}$ is called the **projective linear group** $PGL(2,\mathbb{C})$. Thus we have a group isomorphism

$$PGL(2,\mathbb{C}) \approx \text{MOB}.$$

Checkpoint 3.2.4 Show that the map $\mathcal{T}$ is a group homomorphism. Show that the kernel of $\mathcal{T}$ is

$$\ker \mathcal{T} = \left\{ \begin{bmatrix} k & 0 \\ 0 & k \end{bmatrix}, k \neq 0 \right\}.$$

Homotheties, rotations, translations, and inversions (see Table 3.1.4, p. 51 in Section 3.1, p. 50) are special cases of Möbius transformations. These basic transformations can be viewed as building blocks for general Möbius transformations, as follows.

Proposition 3.2.5 *Every Möbius transformation is a composition of homotheties, rotations, translations, and inversions.*

Checkpoint 3.2.6

1. Identify the values of the coefficients a, b, c, d in a Möbius transformation $z \to \frac{az+b}{cz+d}$ that is a homothety, rotation, translation, and inversion, respectively.

2. Write a Möbius transformation that does "clockwise rotation by one-quarter rotation about the point $2 - i$".

Corollary 3.2.7 *Möbius transformations are conformal.*

Checkpoint 3.2.8 Apply Proposition 3.2.5, p. 56 and Exercise Group 3.1.4.2–5, p. 53 to prove Corollary 3.2.7, p. 56.

3.2.2 The Fundamental Theorem of Möbius Geometry

The Fundamental Theorem of Möbius Geometry says that all three-point sets are congruent. It will turn out to be convenient to work with the "standard" three-point set $\{1, 0, \infty\}$. Given any three distinct complex numbers z_1, z_2, z_3 it is easy to check that the transformation T given by

$$Tz = \frac{z - z_2}{z - z_3} \frac{z_1 - z_3}{z_1 - z_2} \tag{3.2.3}$$

satisfies $Tz_1 = 1, Tz_2 = 0, Tz_3 = \infty$.

Checkpoint 3.2.9 Verify that (3.2.3) satisfies $Tz_1 = 1, Tz_2 = 0, Tz_3 = \infty$.

It is easy to adapt (3.2.3) to the extended complex plane, where one of the z_i may be the point at infinity.

Lemma 3.2.10 Any three-point set is congruent to $\{1, 0, \infty\}$. *Let z_1, z_2, z_3 be distinct points in the extended complex plane. There exists a Möbius transformation T such that $Tz_1 = 1, Tz_2 = 0, Tz_3 = \infty$.*

Checkpoint 3.2.11 Prove Lemma 3.2.10, p. 57 by finding explicit formulas for T for each of the three cases $z_1 = \infty$, $z_2 = \infty$, and $z_3 = \infty$.

(See *Solutions to Exercises* in Appendix B, p. 103.)

To analyze the structure of Möbius transformations, it is useful to consider their **fixed points**, that is, values z such that $Tz = z$. The next Lemma establishes a fact about fixed points that will be used to prove the uniqueness of congruences between three-point sets.

Lemma 3.2.12 *If a Möbius transformation has more than two fixed points, then it is the identity transformation.*

Checkpoint 3.2.13 Prove Lemma 3.2.12, p. 57.

Hint. Solve $z = \frac{az+b}{cz+d}$. You will need to consider cases.

Proposition 3.2.14 The Fundamental Theorem of Möbius Geometry. *A Möbius transformation is completely determined by any three input-output pairs. This means that for any triple of distinct input values z_1, z_2, z_3 in $\hat{\mathbb{C}}$ and any triple of distinct output values w_1, w_2, w_3 in $\hat{\mathbb{C}}$, there is a unique $T \in \mathrm{MOB}$ such that $Tz_i = w_i$ for $i = 1, 2, 3$.*

3.2.3 Cross ratio

Given three distinct points z_1, z_2, z_3 in $\hat{\mathbb{C}}$, we write $(\cdot, z_1, z_2, z_3)$ to denote the unique Möbius transformation that satisfies $z_1 \to 1$, $z_2 \to 0$, and $z_3 \to \infty$ (the existence and uniqueness of this transformation is guaranteed by the Fundamental Theorem of Möbius Geometry). We write (z_0, z_1, z_2, z_3) to denote the image of z_0 under $(\cdot, z_1, z_2, z_3)$. The expression (z_0, z_1, z_2, z_3) is called the **cross ratio** of the 4-tuple z_0, z_1, z_2, z_3. By (3.2.3), we have the following explicit formula for the cross ratio.

$$(z_0, z_1, z_2, z_3) = \frac{z_0 - z_2}{z_0 - z_3} \frac{z_1 - z_3}{z_1 - z_2} \tag{3.2.4}$$

The cross ratio is a useful tool because it is invariant under Möbius transformations. Here is the formal statement with a proof.

Proposition 3.2.15 Cross ratio is invariant. *Let z_1, z_2, z_3 be distinct points in $\hat{\mathbb{C}}$, let $z_0 \in \hat{\mathbb{C}}$, and let T be any Möbius transformation. Then we have*

$$(z_0, z_1, z_2, z_3) = (Tz_0, Tz_1, Tz_2, Tz_3).$$

Proof. The transformations $(\cdot, z_1, z_2, z_3)$ and $(\cdot, Tz_1, Tz_2, Tz_3) \circ T$ both send $z_1 \to 1$, $z_2 \to 0$, and $z_3 \to \infty$, so they must be equal, by the Fundamental Theorem. Now apply both transformations to z_0.   ∎

3.2.4 Clines (generalized circles)

A Euclidean circle or straight line is called a **cline** (pronounced "kline") or **generalized circle**. The propositions and corollaries in this subsection show that the set of all clines is a single congruence class of figures in Möbius geometry.

Proposition 3.2.16 *Let z_1, z_2, z_3 be distinct points in $\hat{\mathbb{C}}$, let $T = (\cdot, z_1, z_2, z_3)$, and let $C_T = T^{-1}\left(\hat{\mathbb{R}}\right)$ be the inverse image of the extended real line $\hat{\mathbb{R}} = \mathbb{R} \cup \{\infty\}$ under T. Then C_T is a Euclidean circle or straight line. Furthermore, C_T is the unique Euclidean circle or straight line that contains the points z_1, z_2, z_3.*

Corollary 3.2.17 *The cross ratio (z_0, z_1, z_2, z_3) is real if and only if z_0, z_1, z_2, z_3 lie on a Euclidean circle or straight line.*

Checkpoint 3.2.18 Explain how Corollary 3.2.17, p. 58 follows from Proposition 3.2.16, p. 58.

Corollary 3.2.19 *Let C be a Euclidean circle or straight line in $\hat{\mathbb{C}}$ and let T be any Möbius transformation. Then $T(C)$ is a Euclidean circle or straight line.*

Checkpoint 3.2.20 Explain how Corollary 3.2.19, p. 58 follows from Proposition 3.2.16, p. 58.

Hint. Let z_1, z_2, z_3 be three points on C and let $w_1 = Tz_1, w_2 = Tz_2, w_3 = Tz_3$. Let $U = (\cdot, z_1, z_2, z_3)$ and let $V = (\cdot, w_1, w_2, w_3)$. Explain why $T = V^{-1}U$ and hence that $T(C) = V^{-1}U(C) = V^{-1}(\hat{\mathbb{R}})$ must be a cline.

Corollary 3.2.21 *All clines are congruent in Möbius geometry.*

Checkpoint 3.2.22 Explain how Corollary 3.2.21, p. 58 follows from Proposition 3.2.16, p. 58.

Hint. Start here: Let C, D be two clines. Let z_1, z_2, z_3 be three points on C and let w_1, w_2, w_3 be three points on D. Let $U = (\cdot, z_1, z_2, z_3)$, let $V = (\cdot, w_1, w_2, w_3)$, and let $T = V^{-1}U$. Explain why T takes C to D.

3.2.5 Symmetry with respect to a cline

Geometrically, the conjugation map $z \to z^*$ in the complex plane is *reflection* across the real line. This "mirror" symmetry generalizes to symmetry with respect to any cline, as follows. Given a cline C that contains z_1, z_2, z_3 in $\hat{\mathbb{C}}$, let $T = (\cdot, z_1, z_2, z_3)$. Given any point z, the **symmetric point with respect to** C is

$$z^{*C} = (T^{-1} \circ \text{conj} \circ T)(z) \qquad (3.2.5)$$

where $\text{conj} \colon \hat{\mathbb{C}} \to \hat{\mathbb{C}}$ is the extension of the conjugation map to the extended complex plane that sends $\infty \to \infty^* = \infty$. The idea is to map C to the real line via T, then conjugate, then map the real line back to C. See Figure 3.2.23, p. 59.

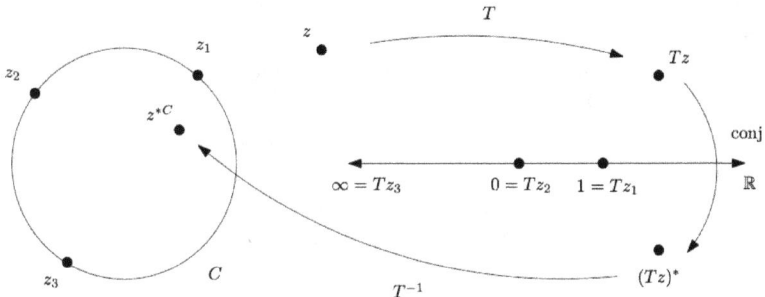

Figure 3.2.23 Symmetric points z, z^{*C} with respect to the circle C

Checkpoint 3.2.24 Show that the symmetric point of a symmetric point is the point you started with. That is, show that $(z^{*C})^{*C} = z$. This allows us to speak of "a pair of symmetric points" without ambiguity.

Hint. $(T^{-1} \circ \text{conj} \circ T) \circ (T^{-1} \circ \text{conj} \circ T) = \text{Id}$

Proposition 3.2.25 *The definition (3.2.5) of z^{*C} depends only on the cline C, and not on the three points z_1, z_2, z_3.*

Proposition 3.2.26 *Let C be a cline and let S be a Möbius transformation. If z, z' are a pair of points that are symmetric with respect to C, then Sz, Sz' are symmetric with respect to the cline $S(C)$. That is, we have*

$$(Sz)^{*S(C)} = S(z^{*C}).$$

Proof. Let z_1, z_2, z_3 be three points on C, so that Sz_1, Sz_2, Sz_3 are three points on $S(C)$. Let $T = (\cdot, z_1, z_2, z_3)$ and let $U = (\cdot, Sz_1, Sz_2, Sz_3)$. By invariance of the cross ratio, we have

$$(U \circ S)z = Tz.$$

Thus we have

$$(Sz)^{*S(C)} = (U^{-1} \circ \text{conj} \circ U)(Sz) \quad \text{(by definition)}$$

$$= (S \circ S^{-1} \circ U^{-1} \circ \operatorname{conj} \circ U \circ S)(z)$$
$$= S(S^{-1} \circ U^{-1} \circ \operatorname{conj} \circ U \circ S)(z)$$
$$= S(T^{-1} \circ \operatorname{conj} \circ T)(z)$$
$$= S(z^{*C})$$

as desired.   ■

3.2.6 Normal forms

We conclude this section on Möbius geometry with a discussion of the **normal form** of a Möbius transformation. We begin with a Lemma.

Lemma 3.2.27 *If a Möbius transformation has exactly two fixed points* 0 *and* ∞, *then it has the form* $z \to \alpha z$ *for some nonzero* $\alpha \in \mathbb{C}$. *If a Möbius transformation has a single fixed point at* ∞, *then it has the form* $z \to z + \beta$ *for some nonzero* $\beta \in \mathbb{C}$.

Now suppose that a Möbius transformation T has two fixed points, p and q. Let S be given by $Sz = \frac{z-p}{z-q}$. Let $w = Sz$ and let $U = S \circ T \circ S^{-1}$ be the transformation of the w-plane that is conjugate to T via S (see Exercise Group 1.3.4.3–6, p. 11). It is easy to check that U has exactly two fixed points 0 and ∞. Applying the previous Lemma, we have $Uw = \alpha w$ for some nonzero $\alpha \in \mathbb{C}$. Applying both sides of $S \circ T = U \circ S$ to z, we have the following *normal form* for T.

$$\frac{Tz - p}{Tz - q} = \alpha \frac{z - p}{z - q} \tag{3.2.6}$$

The transformation T is called **elliptic**, **hyperbolic**, or **loxodromic** if U is a rotation ($|\alpha| = 1$), a homothety ($\alpha > 0$), or neither, respectively.

Finally, suppose that a Möbius transformation T has exactly one fixed point at p. Let S be given by $Sz = \frac{1}{z-p}$. Again, let $w = Sz$ and let $U = S \circ T \circ S^{-1}$. This time, U has exactly one fixed point at ∞. Applying the Lemma, we have $Uw = w + \beta$ for some nonzero $\beta \in \mathbb{C}$. Applying both sides of $S \circ T = U \circ S$ to z, we have the following *normal form* for T.

$$\frac{1}{Tz - p} = \frac{1}{z - p} + \beta \tag{3.2.7}$$

A Möbius transformation of this type is called **parabolic**. Here is a summary of the classification terminology associated with normal forms.

Table 3.2.28 Summary of normal forms of $T \in$ MOB

normal form type	normal form	conjugate transformation type
elliptic	$\frac{Tz-p}{Tz-q} = \alpha \frac{z-p}{z-q}, |\alpha| = 1$	rotation
hyperbolic	$\frac{Tz-p}{Tz-q} = \alpha \frac{z-p}{z-q}, \alpha > 0$	homothety
loxodromic	$\frac{Tz-p}{Tz-q} = \alpha \frac{z-p}{z-q}, \alpha \neq 0$ $|\alpha| \neq 1, \alpha \not> 0$	composition of homothety with rotation
parabolic	$\frac{1}{Tz-p} = \frac{1}{z-p} + \beta, \beta \neq 0$	translation

3.2.7 Steiner circles

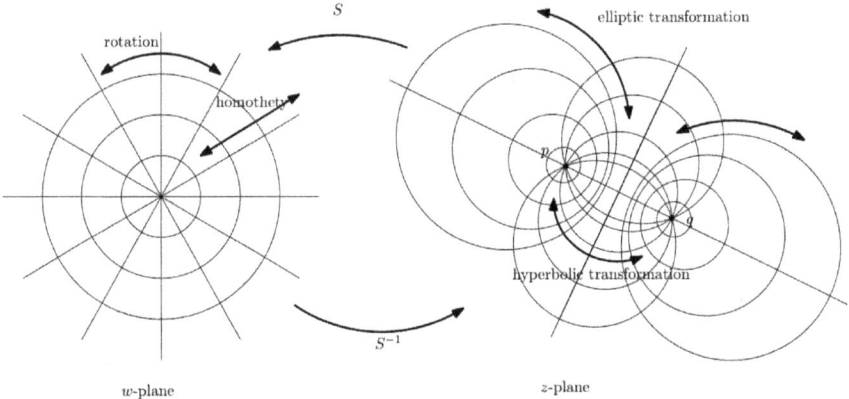

Figure 3.2.29 The polar coordinate grid and Steiner circle coordinate grid

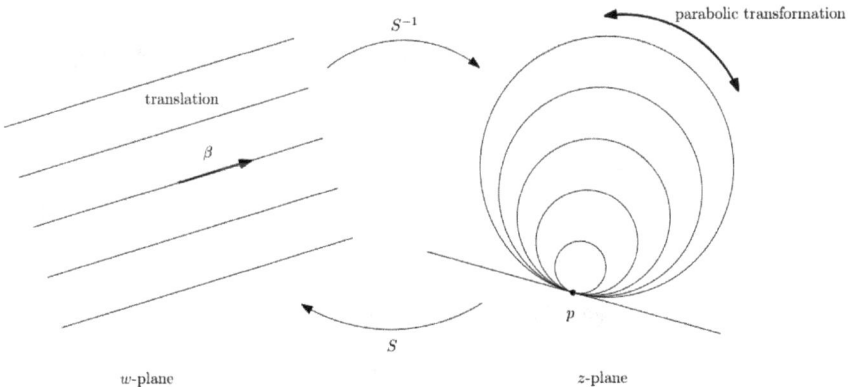

Figure 3.2.30 Degenerate coordinate grid lines and degenerate Steiner circles

This discussion of normal forms (above) shows that any non-identity Möbius transformation is conjugate to one of two basic forms, namely,

$w \to \alpha w$ or $w \to w + \beta$. The natural coordinate system for depicting the action of $w \to \alpha w$ is standard polar coordinates. See Figure 3.2.29, p. 61. A homothety is a flow along radial lines and a rotation is a flow around polar circles. The natural "degenerate" coordinate system for depicting a translation $w \to w + \beta$ is a family of lines parallel to the line that contains the origin and β. A translation by β is a flow along these parallel lines. See Figure 3.2.30, p. 61.

Pulling the polar and degenerate coordinate grids back to the z-plane by S^{-1} leads to coordinate grids called **Steiner circles**.[1] In the case where T has two fixed points p, q, the conjugating map $Sz = \frac{z-p}{z-q}$ takes $p \to 0$, $q \to \infty$. Therefore S^{-1} maps $0 \to p$ and $\infty \to q$. The transformation S^{-1} maps radial lines in the w-plane to clines in the z-plane that contain p and q called **Steiner circles of the first kind** and S^{-1} maps polar circles in the w-plane to clines in the z-plane called **Steiner circles of the second kind** or **circles of Apollonius**. See Figure 3.2.29, p. 61.

In the case where T has one fixed point p, the conjugating map $Sz = \frac{1}{z-p}$ sends $p \to \infty$, so S^{-1} maps $\infty \to p$, and S^{-1} maps lines in the w-plane that are parallel to the line through 0 and β to clines in the z-plane that contain p. Every cline in this family is tangent to every other cline in this family at exactly the one point p. Clines in this family are called **degenerate Steiner circles**. See Figure 3.2.30, p. 61. Table 3.2.31, p. 62 summarizes the graphical depiction of Möbius transformations.

Table 3.2.31 Summary of Steiner circle pictures of Möbius transformations

normal form type	graphical dynamic
elliptic	flow along Steiner circles of the second kind
hyperbolic	flow along Steiner circles of the first kind
loxodromic	composition of elliptic and hyperbolic flows
parabolic	flow along degenerate Steiner circles

3.2.8 Exercises

1. **Decomposition of Möbius transformations into four basic types.**

   (a) Let a be a nonzero complex constant. Explain why the transformation $z \to az$ is a composition of a homothety and a rotation.

   (b) Explicitly identify each homothety, rotation, translation, and inversion in (3.2.8) to (3.2.11) in the derivation below for the case $c \neq 0$.

   $$z \to cz + d \tag{3.2.8}$$

---

[1]The convention for which Steiner circles are considered "first" or "second" kinds is not universal. Here we follow the convention used by Ahlfors [1] and Henle [4].

$$\rightarrow \frac{1}{cz+d} \tag{3.2.9}$$

$$\rightarrow \frac{bc-ad}{cz+d}+a \tag{3.2.10}$$

$$\rightarrow \frac{1}{c}\left(\frac{bc-ad}{cz+d}+a\right) \tag{3.2.11}$$

$$=\frac{az+b}{cz+d} \tag{3.2.12}$$

(c) Write your own decomposition for the case $c=0$.

2. Prove the Fundamental Theorem of Möbius Geometry (Proposition 3.2.14, p. 57).

   Hint. Start by applying Lemma 3.2.10, p. 57 to the triples z_1, z_2, z_3 and w_1, w_2, w_3.

3. Find Möbius transformations that make the following assignments.

   (a) $1 \rightarrow a, 0 \rightarrow b, \infty \rightarrow c$

   (b) $a \rightarrow d, b \rightarrow e, c \rightarrow f$

4. Prove Proposition 3.2.16, p. 58. Suggestion: Let $Tz = \frac{az+b}{cz+d}$, then manipulate $Tz = (Tz)^*$ to an equation with $|z|^2, z, z^*$ terms and coefficients involving a, b, c, d and their conjugates. For the case when the coefficient of $|z|^2$ is not zero, use "complex completing the square" (see (1.1.3)) to derive the equation of a circle. Peek at the first part hint below if you need to, and use it to work partially forwards from $Tz = (Tz)^*$, and partially backwards from the equation in the hint. For the case when the coefficient of $|z|^2$ is zero, derive the equation for a line. For the "furthermore" statement, you will need to show that any three points in the plane determine a unique circle or straight line. This is a Euclidean statement, and it is straightforward to prove this using Euclidean methods (peek at the second part of the hint below if you need to).

   Hint. Equation for the circle:

$$\left| z - \left(\frac{a^*d-bc^*}{ac^*-a^*c}\right)\right|^2 = \left|\frac{ad-bc}{ac^*-a^*c}\right|^2$$

   Why do three noncollinear points P, Q, R determine a unique circle? The center C of the circle must be the intersection of the perpendicular bisectors of segments PQ, QR. The radius must be the distance from C to any one of P, Q, R.

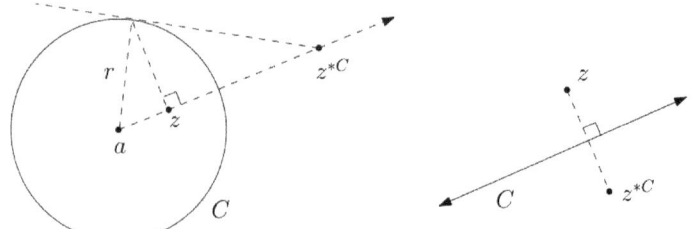

Figure 3.2.32 Geometry for symmetric points z, z^{*C} with respect to the cline C

5.   **Symmetry with respect to a cline.**

(a) Prove Proposition 3.2.25, p. 59 by completing the details in the following outline. Let $C = T^{-1}(\hat{\mathbb{R}})$ be a cline that contains points z_1, z_2, z_3, where $T = (\cdot, z_1, z_2, z_3)$. First, consider the case when C is a circle, say, with equation $|z - a| = r$, where a, r are the center and radius, respectively, of C (see (1.1.1)).

  i. Square both sides of $|z - a| = r$ and solve for z^* to get

$$z^* = \frac{r^2}{z - a} + a^*.$$

  ii. We have

$$(\text{conj} \circ T)(z) = (z^*, z_1^*, z_2^*, z_3^*)$$

$$= \left(z^*, \frac{r^2}{z_1 - a} + a^*, \frac{r^2}{z_2 - a} + a^*, \frac{r^2}{z_3 - a} + a^* \right)$$

$$\overset{(1)}{=} \left(\frac{r^2}{z^* - a^*} + a, z_1, z_2, z_3 \right)$$

$$= T \left(\frac{r^2}{(z - a)^*} + a \right).$$

Explain how invariance of the cross ratio is used to justify the equality (1) in this derivation.

  iii. Explain how we can conclude that $z^{*C} = \frac{r^2}{(z-a)^*} + a$, and that indeed this does not depend on the choice of z_1, z_2, z_3.

  iv. Now suppose that C is a line, say, with equation $\text{Im}(az+b) = 0$ (see (1.1.2)). This is the same as $az + b = (az + b)^*$. Solve to get

$$z^* = \frac{az + b - b^*}{a^*}.$$

  v. We have

$$(\text{conj} \circ T)(z) = (z^*, z_1^*, z_2^*, z_3^*)$$

$$= \left(z^*, \frac{az_1 + b - b^*}{a^*}, \frac{az_2 + b - b^*}{a^*}, \frac{az_3 + b - b^*}{a^*} \right)$$

$$\overset{(1)}{=} \left(\frac{z^* a^* + b^* - b}{a}, z_1, z_2, z_3 \right)$$

$$= T \left(\frac{z^* a^* + b^* - b}{a} \right).$$

Explain how invariance of the cross ratio is used to justify the equality (1) in this derivation.

vi. Explain how we can conclude that $z^{*C} = \frac{z^* a^* + b^* - b}{a}$, and that indeed this does not depend on the choice of z_1, z_2, z_3.

(b) Show that z, z' are symmetric with respect to a circle C with center a and radius r if and only if $|z - a||z' - a| = r^2$ and z, z' lie on the same ray emanating from a. See Figure 3.2.32, p. 64.

(c) Show that z, z' are symmetric with respect to a line C if and only if z, z' are reflections of one another across C. See Figure 3.2.32, p. 64.

Hint. For part (b): The equation for z^{*C} in part iii gives $(z' - a)(z - a)^* = r^2$. For the first statement, take the norm of both sides. For the second statement, take the argument of both sides.

Normal forms and Steiner circles.

6.   Prove Lemma 3.2.27, p. 60.

7.   Find the normal form and sketch a graph using Steiner circles for the following transformations.

   (a) $z \to \dfrac{1}{z}$

   (b) $z \to \dfrac{3z - 1}{z + 1}$

8.   Let p be the single fixed point of a Möbius transformation that is conjugate to $w \to w + \beta$ via $Sz = \frac{1}{z-p}$. Show that the single line in the degenerate Steiner clines through p is parallel to the direction given by β^*.

   Hint. Show that $S^{-1}w = \frac{pw+1}{w}$, so S^{-1} takes $0, \beta, \infty$ to $\infty, \frac{p\beta+1}{\beta}, p$. Thus the single degenerate Steiner straight line through p is in the direction given by the argument of $\frac{p\beta+1}{\beta} - p = \frac{1}{\beta}$, that is, in the direction $-\arg(\beta)$.

9.   Show that a (generalized) circle of Apollonius (a Steiner circle of

the second kind) is characterized as the set of points of the form

$$C = \left\{ P \in \mathbb{C} : \frac{d(P, A)}{d(P, B)} = k \right\}$$

for some $A, B \in \mathbb{C}$ and some real constant $k > 0$.

3.3 Hyperbolic geometry

Before the discovery of hyperbolic geometry, it was believed that Euclidean geometry was the only possible geometry of the plane. In fact, hyperbolic geometry arose as a byproduct of efforts to prove that there was no alternative to Euclidean geometry. In this section, we present a Kleinian version of hyperbolic geometry.

Definition 3.3.1 Let $\mathbb{D} = \{z \colon |z| < 1\}$ denote the open unit disk in the complex plane. The **hyperbolic group**, denoted HYP, is the subgroup of the Möbius group MOB of transformations that map $\mathbb{D}$ onto itself. The pair $(\mathbb{D}, \text{HYP})$ is the **(Poincaré) disk model** of hyperbolic geometry. ◊

Comments on terminology: Beware of the two different meanings of the adjective "hyperbolic". To say that a Möbius transformation is hyperbolic means that it is conjugate to a homothety (see Subsection 3.2.6, p. 60). That is not the same thing as an element of the group of hyperbolic transformations.

3.3.1 The hyperbolic transformation group

Our first task is to characterize transformations in the group HYP. We begin with an observation about Möbius transformations that map one "side" of a cline to itself. This is pertinent because the disk $\mathbb{D}$ is the "inside" of the cline which is the unit circle. It will be useful to start with a general case.

Any cline C divides the extended plane into two regions. If C is a Euclidean circle, we might called these regions the "inside" and the "outside" of C. If C is a Euclidean straight line, we simply have one side and the other of C.

Proposition 3.3.2 Möbius transformations that map one side of a cline to itself. *Let C be a cline, and let D, E be the two disjoint regions of $\hat{\mathbb{C}} \setminus C$. Let T be a Möbius transformation that maps D onto itself. Then T also maps E onto itself, and T maps C onto itself.*

Proof. Sketch: Suppose that T maps D onto itself. The "other side" of C is the set of points that are symmetric, with respect to C, to the points in D. By Proposition 3.2.26, p. 59, T maps symmetric points to symmetric points, so T maps E into itself. It is easy to check that, in fact, T maps E onto itself. By elimination, it must be that T maps C onto itself. ∎

Checkpoint 3.3.3 Complete the sketch of the proof of Proposition 3.3.2, p. 67.

Corollary 3.3.4 *If $T \in \text{HYP}$, then T maps the unit circle onto itself.*

Given $T \in \text{HYP}$, let $z_0 \in \mathbb{D}$ be the point that T maps to 0. It must be that T maps the symmetric point $1/z_0^*$ to ∞. Let z_1 be the point that T

maps to 1. Then T has the form (see (3.2.3))

$$Tz = \frac{z - z_0}{z - 1/z_0^*} \frac{z_1 - 1/z_0^*}{z_1 - z_0}.$$

Multiplying top and bottom by $-z_0^*$, and setting $\alpha = -z_0^* \frac{z_1 - 1/z_0^*}{z_1 - z_0}$, we have

$$Tz = \alpha \frac{z - z_0}{1 - z_0^* z}.$$

A straightforward derivation shows that $|\alpha| = 1$, so that we have (3.3.1) below. Another computation establishes an alternative formula (3.3.2) for $T \in \text{HYP}$. See Exercise 3.3.6.1, p. 74.

Proposition 3.3.5 *A Möbius transformation T is in* HYP *if and only if T can be written in the form*

$$Tz = e^{it} \frac{z - z_0}{1 - z_0^* z} \qquad (3.3.1)$$

for some $t \in \mathbb{R}$ and $z_0 \in \mathbb{D}$. Alternatively, we have $T \in$ HYP if and only if T can be written in the form

$$Tz = \frac{az + b}{b^* z + a^*} \qquad (3.3.2)$$

for some $a, b \in \mathbb{C}$ such that $|a|^2 - |b|^2 = 1$.

3.3.2 Classification of clines in hyperbolic geometry

The clines of Möbius geometry are classified into several types in hyperbolic geometry, as summarized in Table 3.3.6, p. 68.

Table 3.3.6 Clines in hyperbolic geometry

hyperbolic curve type	cline type
hyperbolic straight line	a cline that intersects the unit circle at right angles
hyperbolic circle	a circle completely contained in $\mathbb{D}$
horocycle	a circle with all but one point in $\mathbb{D}$, tangent to the unit circle
hypercycle	a cline that intersects the unit circle at a non-right angle

Checkpoint 3.3.7 Show that each of the four categories of clines in Table 3.3.6, p. 68 is preserved by transformations in the hyperbolic group. That is, show that any transformation in the hyperbolic group takes hyperbolic straight lines to hyperbolic straight lines, takes hyperbolic circles to hyperbolic circles, takes horocycles to horocycles, and takes hypercycles to

hypercycles.

Checkpoint 3.3.8 Show that a hyperbolic straight line that contains 0 must be a diameter of the unit circle.

Hint. Prove the contrapositive: assume C is a hyperbolic straight line that is also a Euclidean circle, and intersects the unit circle orthogonally at p. Give an argument why C *cannot* contain 0.

Checkpoint 3.3.9 Show that all hyperbolic straight lines are congruent.

Hint. Start by showing that any hyperbolic straight line is congruent to $\hat{\mathbb{R}}$.

3.3.3 Normal forms for the hyperbolic group

In this subsection, we follow the development of normal forms for general Möbius transformations given in Subsection 3.2.6, p. 60 to derive normal forms and graphical interpretations for transformations in the hyperbolic group. We begin with an observation about fixed points of a Möbius transformation that maps a cline to itself.

Lemma 3.3.10 *Let $T \in$ MOB and let C be a cline. If $Tz = z$, then $T(z^{*C}) = z^{*C}$.*

Proof. Apply Proposition 3.2.26, p. 59. ∎

Now let T be a non-identity element of HYP. The fact that T maps the unit circle to itself implies that there are exactly three possible cases for fixed points of T.

1. There is a pair of fixed points p, q with $|p| < 1$, $|q| > 1$, and $q = \frac{1}{p^*}$, that is, p, q are a pair of symmetric points (with respect to the unit circle) that do not lie on the unit circle.

2. There is a pair of fixed points that lie on the unit circle.

3. There is a single fixed point that lies on the unit circle.

Checkpoint 3.3.11 Give an argument to justify the three cases above.

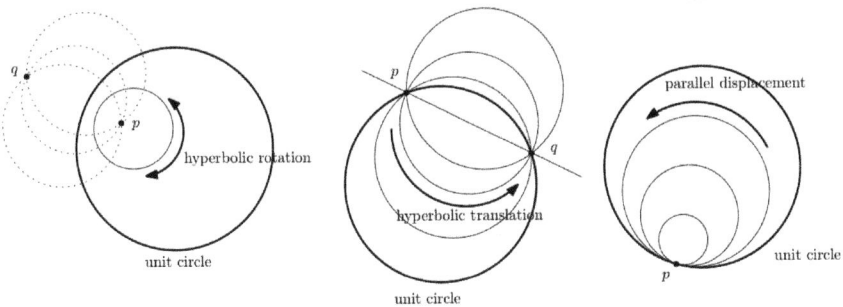

Figure 3.3.12 Three types of hyperbolic transformations

For cases 1 and 2 above, the map T acting on the z-plane is conjugate to the map $U = S \circ T \circ S^{-1}$ acting on the w-plane by $Uw = \lambda w$, for some nonzero $\lambda \in \mathbb{C}$, via the map $w = Sz = \frac{z-p}{z-q}$. In case 1, the map S takes the unit circle to some polar circle, say C, so U must map C to itself. It follows that $|\lambda| = 1$, so the Möbius normal form type for T is elliptic. The action of T is a rotation about Steiner circles of the second kind (hyperbolic circles) with respect to the fixed points p, q. A transformation $T \in \mathrm{HYP}$ of this type is called a **hyperbolic rotation**. See Figure 3.3.12, p. 69.

For case 2, the map $w = Sz = \frac{z-p}{z-q}$ takes the unit circle to a straight line, say L, through the origin, so $U = S \circ T \circ S^{-1}$ must map L to itself. It follows that λ is real. Since S maps $\mathbb{D}$ to one of the two half planes on either side of L, the map U must take this half plane to itself. If follows that λ must be a *positive* real number, so the Möbius normal form type for T is hyperbolic. The action of T is a flow about Steiner circles of the first kind (hypercycles and one hyperbolic straight line) with respect to the fixed points p, q. A transformation $T \in \mathrm{HYP}$ of this type is called a **hyperbolic translation**. See Figure 3.3.12, p. 69.

For case 3, the conjugating map $w = Sz = \frac{1}{z-p}$ takes T to $U = S \circ T \circ S^{-1}$ of the form $Uw = w + \beta$ for some $\beta \neq 0$. The Möbius normal form type for T is parabolic. The action of T is a flow along degenerate Steiner circles (horocycles) tangent to the unit circle at p. A transformation $T \in \mathrm{HYP}$ of this type is called a **parallel displacement**. See Figure 3.3.12, p. 69.

This completes the list of transformation types for the hyperbolic group. See Table 3.3.13, p. 70 for a summary.

Table 3.3.13 Normal forms for the hyperbolic group

hyperbolic transformation type	Möbius normal form	graphical dynamic
hyperbolic rotation	elliptic	flow around hyperbolic circles
parallel displacement	parabolic	flow around horocycles
hyperbolic translation	hyperbolic	flow along hypercycles
(none)	loxodromic	

3.3.4 Hyperbolic length and area

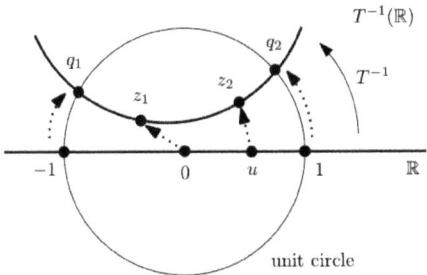

Figure 3.3.14 Constructing the hyperbolic straight line containing two points z_1, z_2

Let z_1, z_2 be distinct points in $\mathbb{D}$. Let $T \in \mathrm{HYP}$ be the transformation that sends $z_1 \to 0$ and $z_2 \to u > 0$. Then $T^{-1}(\mathbb{R})$ is a hyperbolic straight line that contains z_1, z_2. Let $q_1 = T^{-1}(-1)$ and $q_2 = T^{-1}(1)$. See Figure 3.3.14, p. 71.

Checkpoint 3.3.15 Use Proposition 3.3.5, p. 68 to write a formula for the transformation T in the previous paragraph.
(See *Solutions to Exercises* in Appendix B, p. 103.)

A simple calculation verifies that $(0, u, 1, -1) = \frac{1+u}{1-u}$. By invariance of the cross ratio, we have $(z_1, z_2, q_2, q_1) = \frac{1+u}{1-u}$. For $0 \le u < 1$, we have

$$1 \le \frac{1+u}{1-u} < \infty$$

with equality on the left if and only if $u = 0$.

Checkpoint 3.3.16 Do the simple calculation mentioned above. Use (3.2.4).

Now, given any points z_1, z_2 in $\mathbb{D}$, not necessarily distinct, define the quantity $d(z_1, z_2)$ by

$$d(z_1, z_2) = \begin{cases} \ln((z_1, z_2, q_2, q_1)) & z_1 \ne z_2 \\ 0 & z_1 = z_2 \end{cases} \tag{3.3.3}$$

where q_1, q_2 are the ideal points on the hyperbolic straight line connecting z_1, z_2 (with each q_i at the z_i end of the line) as described above, in the case $z_1 \ne z_2$. From the discussion above we have

$$d(z_1, z_2) = \ln\left(\frac{1+u}{1-u}\right) \tag{3.3.4}$$

where $u = \left| \frac{z_2 - z_1}{1 - z_1^* z_2} \right|$.

Checkpoint 3.3.17 Justify the value of u in (3.3.4).

Proposition 3.3.18 *The function d given by (3.3.3) is invariant under the action of the hyperbolic group. That is, we have*

$$d(z_1, z_2) = d(Tz_1, Tz_2) \tag{3.3.5}$$

for all $z_1, z_2 \in \mathbb{D}$ and for all $T \in \text{HYP}$.

Checkpoint 3.3.19 Prove Proposition 3.3.18, p. 72.

The following Proposition shows that d is a *metric* on hyperbolic space, and justifies referring to $d(z_1, z_2)$ as the **(hyperbolic) distance** between the points z_1, z_2.

Proposition 3.3.20 *The function d given by (3.3.3) defines a metric on $\mathbb{D}$. That is, d satisfies the following conditions for all z_1, z_2, z_3 in $\mathbb{D}$.*

1. $d(z_1, z_2) \geq 0$, *and* $d(z_1, z_2) = 0$ *if and only if* $z_1 = z_2$

2. $d(z_1, z_2) = d(z_2, z_1)$

3. $d(z_1, z_3) \leq d(z_1, z_2) + d(z_2, z_3)$ *(the triangle inequality)*

Proof. Property 1 follows immediately from (3.3.4). Property 2 is a simple calculation: just write down the cross ratio expressions for $d(z_1, z_2)$ and $d(z_2, z_1)$ and compare. The proof of Property 3 is outlined in exercise Exercise 3.3.6.4, p. 74. ∎

Now let γ be a curve parameterized by $t \to z(t) = x(t) + iy(t)$, where $x(t), y(t)$ are differentiable real-valued functions of the real parameter t on an interval $a < t < b$. Consider a short segment of γ, say, on an interval $t_0 \leq t \leq t_1$. Let $z_0 = z(t_0)$ and $z_1 = z(t_1)$. Then we have $d(z(t_0), z(t_1)) = \ln\left(\frac{1+u}{1-u}\right)$ where $u = \left|\frac{z_1 - z_0}{1 - z_0^*(z_1)}\right|$. The quantity $|z_1 - z_0|$ is well-approximated by $|z'(t_0)|dt$, where $z'(t) = x'(t) + iy'(t)$ and $dt = t_1 - t_0$. Thus, u is well-approximated by $\frac{|z'(t_0)|}{1 - |z(t_0)|^2} dt$. The first order Taylor approximation for $\ln((1+u)/(1-u))$ is $2u$. Putting this all together, we have the following.

$$\text{Length}(\gamma) = 2 \int_a^b \frac{|z'(t)|}{1 - |z(t)|^2} \, dt \tag{3.3.6}$$

Checkpoint 3.3.21 Show that the first order Taylor approximation of $\ln((1+u)/(1-u))$ is $2u$.

Checkpoint 3.3.22 Find the length of the hyperbolic circle parameterized by $z(t) = \alpha e^{it}$ for $0 \leq t \leq 2\pi$, where $0 < \alpha < 1$.

We conclude this subsection on hyperbolic length and area with an integral formula for the area of a region R in $\mathbb{D}$, following the development in [4]. As a function of the two real variables r and θ, the polar form expression $z = re^{i\theta}$ gives rise to the two parameterized curves $r \to z_1(r) = re^{i\theta}$ (where θ is constant) and $\theta \to z_2(\theta) = re^{i\theta}$ (where r is constant). Using $z_1'(r) = e^{i\theta}$ and $z_2'(\theta) = ire^{i\theta}$, the arc length differential $ds = \frac{2|z'(t)| \, dt}{1 - |z(t)|^2}$ for the two curves are the following.

$$|dz_1| = \frac{2|e^{i\theta}|\, dr}{1 - r^2} = \frac{2\, dr}{1 - r^2}$$

$$|dz_2| = \frac{2|ire^{i\theta}|\, d\theta}{1 - r^2} = \frac{2r\, d\theta}{1 - r^2}$$

Thus we have $dA = \frac{4r\, dr\, d\theta}{(1-r^2)^2}$, so that the area of a region R is

$$\text{Area}(R) = \iint_R dA = \iint_R \frac{4r\, dr\, d\theta}{(1 - r^2)^2}. \tag{3.3.7}$$

Checkpoint 3.3.23 Find the area of the hyperbolic disk $\{|z| \leq \alpha\}$, for $0 < \alpha < 1$.

3.3.5 The upper-half plane model

Definition 3.3.24 The upper half-plane model of hyperbolic geometry. Let $\mathbb{U} = \{z \colon \text{Im}(z) > 0\}$ denote the half of the complex plane above the real axis, and let $\text{HYP}_\mathbb{U}$ denote the subgroup of the Möbius group MOB of transformations that map $\mathbb{U}$ onto itself. The pair $(\mathbb{U}, \text{HYP}_\mathbb{U})$ is the **upper half-plane model** of hyperbolic geometry.   ◇

Proposition 3.3.25 *A Möbius transformation T is in $\text{HYP}_\mathbb{U}$ if and only if T can be written in the form*

$$Tz = \frac{az + b}{cz + d} \tag{3.3.8}$$

such that a, b, c, d are real and $ad - bc > 0$.

Hyperbolic straight lines in the upper half-plane model are clines that intersect the real line at right angles. The hyperbolic distance between two points z_1, z_2 in the upper half-plane is

$$d(z_1, z_2) = \ln((z_1, z_2, q_2, q_1)) \tag{3.3.9}$$

where q_1, q_2 are the points on the (extended) real line at the end of the hyperbolic straight line that contains z_1, z_2, with each q_i on the same "side" as the corresponding z_i. The hyperbolic length of a curve γ parameterized by $t \to z(t) = x(t) + iy(t)$ on the interval $a \leq t \leq b$ is

$$\text{Length}(\gamma) = \int_a^b \frac{|z'(t)|}{y(t)}\, dt. \tag{3.3.10}$$

The hyperbolic area of a region R in $\mathbb{U}$ is

$$\text{Area}(R) = \iint_R dA = \iint_R \frac{dx\, dy}{y^2}. \tag{3.3.11}$$

3.3.6 Exercises

1. Prove Proposition 3.3.5, p. 68 using the following outline.

    (a) Complete the proof of (3.3.1) using this outline: Let $|z| = 1$ and apply Corollary 3.3.4, p. 67. We have

    $$1 = |Tz| = |\alpha| \left| \frac{z - z_0}{1 - z_0^* z} \right|.$$

    Continue this derivation to show that $|\alpha| = 1$.

    (b) Prove (3.3.2) by verifying the following. Given $z_0 \in \mathbb{D}$ and $t \in \mathbb{R}$, show that the assignments $a = \frac{e^{it/2}}{\sqrt{1-|z_0|^2}}, b = \frac{-e^{it/2} z_0}{\sqrt{1-|z_0|^2}}$ satisfy $|a|^2 - |b|^2 = 1$ and that

    $$\frac{az + b}{b^* z + a^*} = e^{it} \frac{z - z_0}{1 - z_0^* z}. \tag{3.3.12}$$

    Conversely, given $a, b \in \mathbb{C}$ with $|a|^2 - |b|^2 = 1$, show that the assignments $t = 2 \arg a$, $z_0 = -\frac{b}{a}$ satisfy $z_0 \in \mathbb{D}$, and that (3.3.12) holds.

2. **Two points determine a line.** Let p, q be distinct points in $\mathbb{D}$. Show that there is a unique hyperbolic straight line that contains p and q.

    Hint. Start by choosing a transformation that sends $p \to 0$. For uniqueness, use Checkpoint 3.3.8, p. 69.

3. **Dropping a perpendicular from a point to a line.** Let L be a hyperbolic straight line and let $p \in \mathbb{D}$ be a point not on L. Show that there is a unique hyperbolic straight line M that contains p and is orthogonal to L.

    Hint. Start by choosing a transformation that sends $p \to 0$. For uniqueness, use Checkpoint 3.3.8, p. 69.

4. **The triangle inequality for the hyperbolic metric.** Show that $d(a, b) \le d(a, c) + d(c, b)$ for all a, b, c in $\mathbb{D}$ using the outline below.

    (a) Show that the triangle inequality holds with strict equality when a, b, c are collinear and c is between a and b. Suggestion: This is a straightforward computation using the cross ratio expressions for the values of d.

    (b) Show that the triangle inequality holds with strict inequality when a, b, c are collinear and c is not between a and b.

    (c) Let $p \in \mathbb{D}$ lie on a hyperbolic line L, let $q \in \mathbb{D}$, let M be a line through q perpendicular to L (this line M exists by Exercise 3.3.6.3, p. 74), and let q' be the point of intersection of L, M.

Show that $d(p, q') \leq d(p, q)$. Suggestion: apply $T \in \text{HYP}$ that takes $p \to 0$ and takes $L \to \mathbb{R}$. Let $t = -\arg(Tq)$ if $\text{Re}(Tq) \geq 0$ and let $t = \pi - \arg(Tq)$ if $\text{Re}(Tq) < 0$. Let $r = e^{it}Tq$. See Figure 3.3.26, p. 75.

(d) Given arbitrary a, b, c, apply a transformation T to send $a \to 0$ and b to a nonnegative real point. Drop a perpendicular from Tc to the real line, say, to c'. Apply results from the previous steps of this outline.

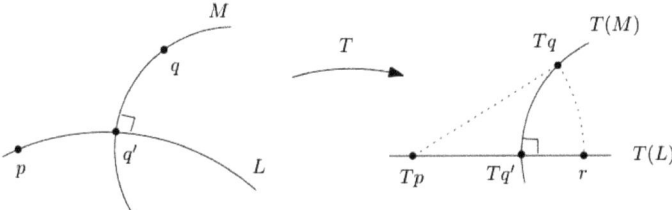

Figure 3.3.26

5. Prove Proposition 3.3.25, p. 73.

    Hint. For the "only if" direction, apply Proposition 3.3.2, p. 67 to the cline $\hat{\mathbb{R}}$, and conclude that T must send the real line to itself. Set z_1, z_2, z_3 to be the preimages under T of $0, 1, \infty$, and then use (3.2.3). For the "if" direction, suppose T has the given form. Let $y > 0$ and show that $\text{Im}(T(x + iy)) > 0$.

6. **Length integral in the upper half-plane model.** This exercise is to establish (3.3.10). The strategy is to obtain the differential expression

$$d(z(t_0), z(t_1)) \approx \frac{|z'(t)|dt}{y(t)}$$

for a curve $z(t) = x(t) + iy(t)$ with $z(t_0) = z_0$, $z(t_1) = z_1$, and $dt = t_1 - t_0$ using the following sequence of steps.

- First, map z_0, z_1 in $\mathbb{U}$ to z_0', z_1' in $\mathbb{D}$ using a transformation μ that preserves distance.

- Using the analysis we used to get the disk model length integral formula (3.3.6), we have

$$d(z_0', z_1') = \ln\left(\frac{1+u}{1-u}\right)$$

where $u = \left|\frac{z_1' - z_0'}{1 - (z_0')^* z_1'}\right|$.

- Translate the above expression in terms of z_0, z_1, and show that the differential approximation is $\frac{|z'(t)|dt}{y(t)}$.

Complete the exercise parts below to carry out the strategy just outlined.

(a) Show that $\mu z = \frac{z-i}{z+i}$ takes $\mathbb{U}$ to $\mathbb{D}$.

(b) Let $z_0' = \mu z_0$ and $z_1' = \mu z_1$. Show that

$$\frac{z_1' - z_0'}{1 - (z_0')^* z_1'} = e^{it} \frac{z_1 - z_0}{z_0^* - z_1}$$

for some real t.

(c) Let $u = \left| \frac{z_1 - z_0}{z_0^* - z_1} \right|$. Show that

$$\ln\left(\frac{1+u}{1-u}\right) \approx \frac{|z(t)|dt}{y(t)}.$$

7. **Area integral in the upper half-plane model.** Adapt the argument in the paragraph preceding the disk model area integral (3.3.7) to establish the upper half-plane area integral (3.3.11).

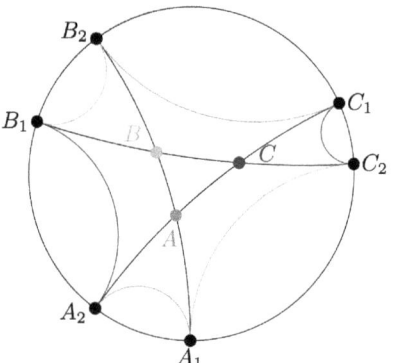

Figure 3.3.27 Hyperbolic triangle $\triangle ABC$

Area of a hyperbolic triangle.   The following sequence of exercises establishes the area formula for hyperbolic triangles.

8. **Area of a doubly-asymptotic triangle.** A triangle with one vertex in $\mathbb{D}$ and two vertices on the unit circle, connected by arcs of circles that are orthogonal to the unit circle, is called a **doubly-asymptotic** hyperbolic triangle. Examples are $\triangle AA_1A_2$, $\triangle BB_1B_2$, and $\triangle CC_1C_2$ in Figure 3.3.27, p. 76.

    (a) Explain why any doubly-asymptotic triangle in the upper half-plane is congruent to the one shown in Figure 3.3.28, p. 77 for some angle α.

    (b) Now use the integration formula for the upper half-plane model to show that the area of the doubly-asymptotic triangle with angle α (at the vertex interior to $\mathbb{U}$) is $\pi - \alpha$.

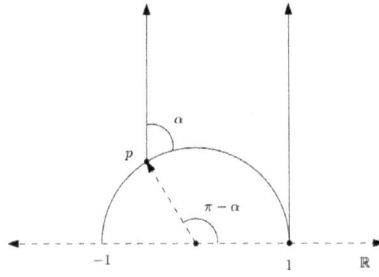

Figure 3.3.28 Doubly-asymptotic hyperbolic triangle in the upper half-plane with vertices $1, p, \infty$ with p on the upper half of the unit circle

9. **Area of an asymptotic n-gon.** A polygon with $n \geq 3$ vertices on the unit circle, connected by arcs of circles that are orthogonal to the unit circle, is called an **asymptotic n-gon.** An example of an asymptotic hexagon is the figure with vertices $A_1, A_2, B_1, B_2, C_2, C_2$ connected by the colored hyperbolic lines in Figure 3.3.27, p. 76. Show that the area of an asymptotic n-gon is $\pi(n - 2)$.

    Hint. Partition the asymptotic n-gon into n doubly-asymptotic triangles.

10. **Area of a hyperbolic triangle.** Let $\triangle ABC$ be a hyperbolic triangle. Extend the three sides AB, BC, AC to six points on the unit circle. See Figure 3.3.27, p. 76. Use a partition of the asymptotic hexagon whose vertices are these six points to show that the area of $\triangle ABC$ is

$$\text{Area}(\triangle ABC) = \pi - (\angle A + \angle B + \angle C). \qquad (3.3.13)$$

    Hint. Partition the asymptotic hexagon with vertices $A_1, A_2, B_1, B_2, C_1, C_2$. Start with the six overlapping doubly-asymptotic triangles whose bases are colored arcs and whose vertex in $\mathbb{D}$ is whichever of A, B, C matches the color of the base. For example, the two red doubly-asymptotic triangles are $\triangle A A_1 A_2$ and $\triangle A C_1 B_2$.

3.4 Elliptic geometry

Elliptic geometry is the geometry of the sphere (the 2-dimensional surface of a 3-dimensional solid ball), where congruence transformations are the rotations of the sphere about its center.

We will work with three models for elliptic geometry: one based on quaternions, one based on rotations of the sphere, and another that is a subgeometry of Möbius geometry. Using the natural identification $xi + yj + zk \leftrightarrow (x, y, z)$ of the pure quaternions $\mathbb{R}_{\mathbb{H}}^3$ with $\mathbb{R}^3$, we will write $S_{\mathbb{H}}^2$ to denote the set of unit pure quaternions, that is,

$$S_{\mathbb{H}}^2 = \{xi + yj + zk \in \mathbb{H} \colon x^2 + y^2 + z^2 = 1\}.$$

We begin by establishing some basic facts about the relevant transformation groups.

3.4.1 The group of unit quaternions

Recall from Section 1.2, p. 5 that $U(\mathbb{H})$ is the set of quaternions of modulus 1. In fact, $U(\mathbb{H})$ is a group.

Checkpoint 3.4.1 Show that $U(\mathbb{H})$ is a group.

Recall that the map $M \colon \mathbb{H} \to \mathcal{M}_{\mathbb{H}}$ sends $r = a + bi + cj + dk$ to the matrix

$$\begin{bmatrix} a + bi & c + di \\ -c + di & a - bi \end{bmatrix}.$$

The image of $U(\mathbb{H})$ under M is the matrix group

$$SU(2) = \left\{ \begin{bmatrix} a & b \\ -b^* & a^* \end{bmatrix} \colon a, b \in \mathbb{C}, |a|^2 + |b|^2 = 1 \right\}, \qquad (3.4.1)$$

called the *special unitary* group. Restricting the domain of M to $U(\mathbb{H})$ and restricting the codomain of M to $M(U(\mathbb{H})) = SU(2)$ is an isomorphism of groups

$$U(\mathbb{H}) \approx SU(2).$$

Checkpoint 3.4.2 Show that $SU(2)$ is a group. Show that $M \colon U(\mathbb{H}) \to SU(2)$ is a homomorphism.

Hint. It is not necessary to perform any new computation to show that M is a homomorphism. Instead, use (1.2.5).

The action of a unit quaternion as a rotation on $\mathbb{R}_{\mathbb{H}}^3$ (see Proposition 1.2.9, p. 8) takes the sphere $S_{\mathbb{H}}^2$ to itself. The action of the group $U(\mathbb{H})$ on $S_{\mathbb{H}}^2$ defines a model of elliptic geometry.

Definition 3.4.3 The **quaternion model** of elliptic geometry is $(S_{\mathbb{H}}^2, U(\mathbb{H}))$. ◊

Checkpoint 3.4.4 Show that the map $U(\mathbb{H}) \to \mathrm{Perm}(S_{\mathbb{H}}^2)$ given by

$$r \to [v \to rvr^*]$$

is a group action.

3.4.2 The group of rotations of the 2-sphere

Let $R_{v,\theta}$ denote the rotation of $\mathbb{R}^3$ about the axis given by a unit vector v by an angle θ. We use the standard orientation, so that a positive value of θ is a counterclockwise rotation of the plane orthogonal to v, as viewed from "above" where v points in the "up" direction. See Figure 3.4.5, p. 79.

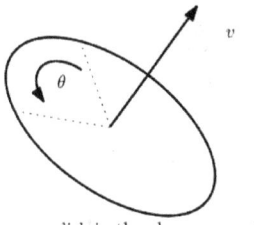

disk in the plane perpendicular to v

Figure 3.4.5 The rotation $R_{v,\theta}$ about the vector v by the angle θ

Notation convention: For readability and convenience, we write $R_{X,\theta}, R_{Y,\theta}, R_{Z,\theta}$ to denote rotations by θ radians about the standard basis vectors $(1,0,0), (0,1,0), (0,0,1)$, respectively.

We will write $\mathrm{Rot}(S^2)$ to denote the set

$$\mathrm{Rot}(S^2) = \{R_{v,\theta} \colon v \in \mathbb{R}^3, |v| = 1, \theta \in \mathbb{R}\}$$

of all rotations. To see why the set $\mathrm{Rot}(S^2)$ is a group[1] under the operation of composition, consider the map $U(\mathbb{H}) \to \mathrm{Rot}(S^2)$ given by $r \to R_r$ established by Proposition 1.2.9, p. 8. The fact that $R_r \circ R_s = R_{rs}$ (see Exercise 1.2.6.2, p. 9) implies that the composition of two rotations is a rotation. The remaining group properties are straightforward. Once we have proved that $\mathrm{Rot}(S^2)$ is a group, the same equation $R_r \circ R_s = R_{rs}$ shows that the map $r \to R_r$ is a homomorphism of groups $U(\mathbb{H}) \to \mathrm{Rot}(S^2)$. The kernel of this homomorphism is $\{\pm 1\}$. This establishes an isomorphism

$$U(\mathbb{H})/\{\pm 1\} \approx \mathrm{Rot}(S^2).$$

Checkpoint 3.4.6 Complete the details to show that $\mathrm{Rot}(S^2)$ is a group. Show that the kernel of the homomorphism $U(\mathbb{H}) \to \mathrm{Rot}(S^2)$ given by $r \to R_r$ is $\{\pm 1\}$.

Hint. Use Proposition 1.2.9, p. 8.

---

[1] For the purpose of a self-contained exposition based on elementary geometry, using only complex and quaternion algebra, we do not utilize the fact that $\mathrm{Rot}(S^2)$ is the group $SO(3)$ of orthogonal transformations of $\mathbb{R}^3$ with determinant 1.

Definition 3.4.7 The **spherical model** of elliptic geometry is $(S^2, \text{Rot}(S^2))$. ◇

We conclude with a useful fact about constructing arbitrary rotations by composing rotations from a specific set elementary types, namely, rotations about the z-axis by arbitrary angles, and rotations about the x-axis by $\pi/2$ radians. We start with a Lemma that shows how to do this for y-axis rotations.

Lemma 3.4.8 Rotations about the y-axis. *An arbitrary rotation about the y-axis is a composition of a rotations about the x-axis by $\pi/2$ radians with a rotation about the z-axis. Specifically, we have the following.*

$$R_{Y,\theta} = R_{X,\pi/2}^{-1} \circ R_{Z,\theta} \circ R_{X,\pi/2} \tag{3.4.2}$$

Proof. Visualize! You can also verify by checking that both sides of (3.4.2) yield the same result when evaluated on the three standard basis vectors. Yet another proof is to do a quaternion computation. ∎

Proposition 3.4.9 Generators for $\text{Rot}(S^2)$**.** *The set*

$$\{R_{Z,\theta} : \theta \in \mathbb{R}\} \cup \{R_{X,\pi/2}\}$$

is a generating set for $\text{Rot}(S^2)$*. This means that any rotation may be written as a composition of rotations about the z-axis and rotations about the x-axis by $\pi/2$ radians.*

Proof. Consider a model of the sphere printed with a map of the world (i.e., a geographic globe) in such a way that the north pole is on top of the sphere and Greenwich, England (at zero degrees longitude) is in the x, z-plane. The sphere in the upper left of Figure 3.4.10, p. 81 depicts this "start" position of the north pole N, Greenwich G, and the great circle C that is the intersection of the sphere with the x, z-plane (C is shown in red in all four spheres for reference). Now let R be an arbitrary rotation. The sphere in the upper right of Figure 3.4.10, p. 81 shows how N, G, and C are transformed by R. The rest of the diagram shows how we can write R^{-1} as a composition of rotations by "putting the north pole back on top" and "putting zero degrees back in place", as follows. From the upper right in the diagram, we "put the north pole back" by first performing a rotation R_{Z,θ_1} about the z-axis that brings the north pole into the x, z-plane. Next, we perform the rotation R_{Y,θ_2} about the y-axis (use the Lemma) to bring the north pole back to the top. Finally, we perform a rotation R_{Z,θ_3} to bring Greenwich back home in the x, z-plane. Reading clockwise from the upper left of the diagram, the sequence of transformations

$$R, R_{Z,\theta_1}, R_{Y,\theta_2}, R_{Z,\theta_3}$$

takes the north pole N through the sequence

$$N \to R(N) \to N' \to N'' = N \to N.$$

Meanwhile, G traces the path

$$G \to R(G) \to G' \to G'' \to G$$

while the great circle C is transformed in the sequence

$$C \to R(C) \to C' \to C'' \to C.$$

This leads to the decomposition $R = R_{Z,-\theta_1} \circ R_{Y,-\theta_2} \circ R_{Z,-\theta_3}$.

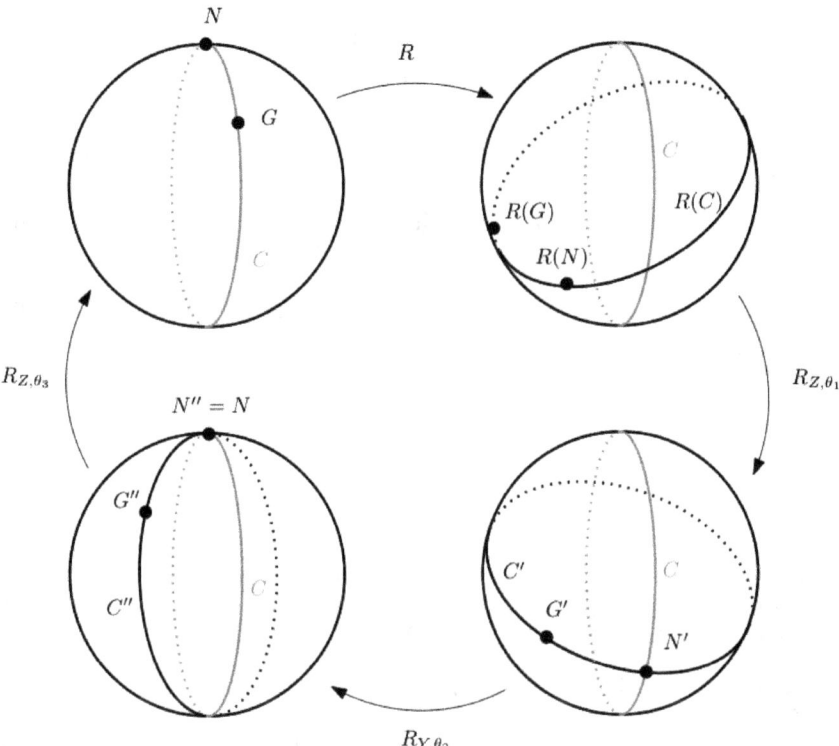

Figure 3.4.10 Decomposition of the rotation R

■

3.4.3 The elliptic subgroup of the Möbius group

Let

$$\text{ELL} = \{s \circ R \circ s^{-1} \ : \ R \in \text{Rot}(S^2)\}$$

denote the group of transformations of $\hat{\mathbb{C}}$ that are conjugate to rotations of S^2 via the stereographic projection s. The group ELL is called the **elliptic group**. It is easy to check that the map $\text{Rot}(S^2) \to \text{ELL}$ given by

$R \to s \circ R \circ s^{-1}$ is an isomorphism of groups, so we have

$$\mathrm{ELL} \approx \mathrm{Rot}(S^2).$$

Checkpoint 3.4.11 Show that ELL is indeed a group. Show that $R \to s \circ R \circ s^{-1}$ is a group isomorphism $\mathrm{Rot}(S^2) \to \mathrm{ELL}$.

Exercises in Section 1.3, p. 10 show that $s \circ R_{Z,\theta} \circ s^{-1}$ is the Möbius transformation $T_{Z,\theta}$ given by $z \to e^{i\theta}z$ (see Exercise 1.3.4.4, p. 12) and that $s \circ R_{X,\pi/2} \circ s^{-1}$ is the Möbius transformation $T_{X,\pi/2}$ given by $z \to \frac{z+i}{iz+1}$ (see Exercise 1.3.4.6, p. 12). The fact (Proposition 3.4.9, p. 80) that rotations of the form $R_{Z,\theta}, R_{X,\pi/2}$ generate $\mathrm{Rot}(S^2)$ implies that the Möbius transformations $T_{Z,\theta}$ and $T_{X,\pi/2}$ generate ELL. Therefore ELL is in fact a subgroup of the Möbius group.

Definition 3.4.12 The **Möbius subgeometry model** of elliptic geometry is $(\hat{\mathbb{C}}, \mathrm{ELL})$. ◇

We can say more about the specific form of elements in ELL in terms of the group homomorphism $\mathcal{T} \colon GL(2,\mathbb{C}) \to \mathrm{MOB}$ that sends the matrix $\begin{bmatrix} a & b \\ c & d \end{bmatrix}$ to the Möbius transformation $z \to \frac{az+b}{cz+d}$ (see (3.2.2)). Observe that the transformations

$$T_{Z,\theta} = \mathcal{T}\left(\begin{bmatrix} e^{i\theta/2} & 0 \\ 0 & e^{-i\theta/2} \end{bmatrix} \right) \tag{3.4.3}$$

$$T_{X,\pi/2} = \mathcal{T}\left(\begin{bmatrix} \frac{1}{\sqrt{2}} & \frac{i}{\sqrt{2}} \\ \frac{i}{\sqrt{2}} & \frac{1}{\sqrt{2}} \end{bmatrix} \right) \tag{3.4.4}$$

are images of elements of the group $SU(2)$ (see (3.4.1)). Because $T_{Z,\theta}, T_{X,\pi/2}$ generate ELL, it follows that every element of ELL is the image under $\mathcal{T}$ of an element of $SU(2)$.

Checkpoint 3.4.13 Let $M_{Z,\theta}, M_{X,\pi/2}$ denote the matrices $\begin{bmatrix} e^{i\theta/2} & 0 \\ 0 & e^{-i\theta/2} \end{bmatrix}, \begin{bmatrix} 1/\sqrt{2} & i/\sqrt{2} \\ i/\sqrt{2} & 1/\sqrt{2} \end{bmatrix}$, respectively. Verify that $M_{Z,\theta}, M_{X,\pi/2}$ are indeed elements of $SU(2)$. Verify (3.4.3) and (3.4.4). Explain the final comment in the paragraph above. Why does it follow that every element of ELL is the image of an element of $SU(2)$?

Thus we have proved the following explicit formula for elements of ELL.

Proposition 3.4.14 Formula for transformations in the elliptic group. *A map T is an element of* ELL *if and only if T may be written in the form $Tz = \frac{az+b}{-b^*z+a^*}$ for some $a, b \in \mathbb{C}$ with $|a|^2 + |b|^2 = 1$.*

Setting $e^{i\theta} = \frac{a}{a^*}$ and $z_0 = \frac{-b}{a}$, we have the following alternative form for $T \in$ ELL.

$$Tz = e^{i\theta}\frac{z - z_0}{z_0^* z + 1} \tag{3.4.5}$$

3.4.4 Circles in S^2 and clines in $\hat{\mathbb{C}}$

A *circle* in S^2 is a circle in a plane intersecting S^2. A **great circle** is the intersection of S^2 with a plane through the origin. In elliptic geometry, a great circle is called an **elliptic straight line** because the path of shortest length connecting two given points in S^2 is an arc of a great circle. Circles in S^2 that are not great circles are called **elliptic cycles**. Elliptic straight lines and elliptic cycles in the Möbius subgeometry model $(\hat{\mathbb{C}}, \text{ELL})$ are stereographic projections of elliptic straight lines and elliptic cycles in the spherical model. It turns out that elliptic straight lines and elliptic cycles in $\hat{\mathbb{C}}$ are in fact clines. Here is the statement and proof.

Proposition 3.4.15 Stereographic projection takes circles to clines.
Let C be a circle that is the intersection of S^2 with a plane in $\mathbb{R}^3$. If C contains the north pole $(0,0,1)$ of S^2, then $s(C \setminus \{(0,0,1)\})$ is a Euclidean straight line in $\mathbb{C}$. Otherwise, $s(C)$ is a circle in $\mathbb{C}$.

Proof. Proof sketch: The statement about the case when C contains the north pole is geometrically clear. For the case when C does not contain $(0,0,1)$, choose a rotation R of S^2 that takes some point on C to the north pole. Again, let $T = s \circ R \circ s^{-1}$ be the conjugate element in ELL. It is clear that R takes C to a circle, that s takes $R(C)$ to a Euclidean straight line, and that T^{-1} takes $s(R(C))$ to a cline (because T^{-1} is a Möbius transformation!). Thus $s(C) = (T^{-1} \circ s \circ R)(C)$ is a cline. Because $(0,0,1)$ is not on C, it must be that ∞ is not on $s(C)$, so $s(C)$ is a circle in $\mathbb{C}$. ∎

3.4.5 Angles and orientation on S^2

The standard orientation for angles on S^2 (see Subsection 3.4.2, p. 79) is also called the *outward-pointing normal* orientation. The standard orientation measures angles from the viewpoint of an observer standing on the *outside* of the sphere. The *inward-pointing normal* orientation is the reverse orientation that measures angles from the viewpoint of an observer walking on the inside of the sphere. See Figure 3.4.16, p. 84.

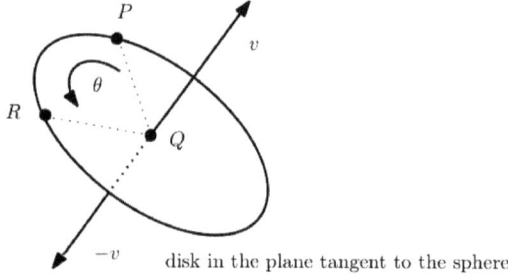

Figure 3.4.16 Two orientations on the sphere: oriented angle $\angle PQR$ is positive viewed from the outward-pointing normal vector v, but is negative viewed from the inward-pointing vector $-v$.

Corollary 3.4.17 Stereographic projection is conformal. *Stereographic projection preserves oriented angles with respect to the inward-pointing normal orientation.*

Proof. Proof sketch: Begin with the special case of curves C_1, C_2 that intersect at the south pole $S = (0, 0, -1)$. The lines L_1, L_2 that are tangent to C_1, C_2 at S lie in planes Π_1, Π_2 that contain the south pole and the origin. The tangents L_1, L_2 also lie in the plane $z = -1$ tangent to the sphere at the south pole. It is clear that the lines L_1', L_2' tangent to $s(C_1), s(C_2)$ at $s(S) = 0$ are straight lines that intersect at the origin. The angle made by L_1, L_2 is the same as the angle made by the planes Π_1, Π_2, but with inward-normal orientation! See Figure 3.4.18, p. 84. Now suppose two curves intersect at P. Choose a rotation R of S^2 that takes P to the south pole, and let $T = s \circ R \circ s^{-1}$ be the conjugate element in ELL. It is clear that R and T are conformal (because T is a Möbius transformation!). Now the fact that s is conformal at $(0, 0, -1)$ implies that $s = T^{-1} \circ s \circ R$ is conformal at P.

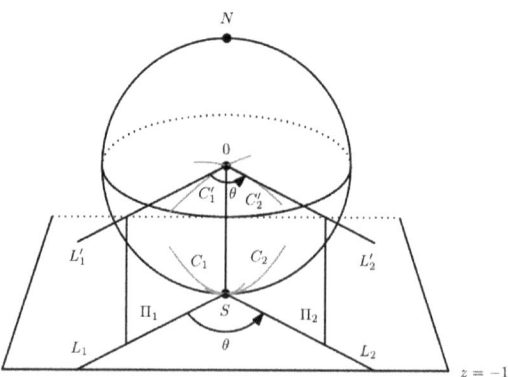

Figure 3.4.18 Stereographic projection is conformal at $S = (0, 0, -1)$.

■

3.4.6 Elliptic length and area

The distance between points P, Q on S^2 is the length of the arc of a great circle that connects them. Because the sphere has radius 1, the arc length is the same as the radian measure of the angle $\angle POQ$, where O is the origin. From vector calculus, we have the following dot product formula.

$$(\overrightarrow{OP}) \cdot (\overrightarrow{OQ}) = |\overrightarrow{OP}||\overrightarrow{OQ}| \cos(\angle POQ)$$

Solving for $\cos(\angle POQ)$, we obtain the formula for the distance $d_{S^2}(P, Q)$ between points P, Q in S^2.

$$d_{S^2}(P, Q) = \cos^{-1}\left((\overrightarrow{OP}) \cdot (\overrightarrow{OQ})\right) \qquad (3.4.6)$$

To "transfer" the metric (3.4.6) to $\hat{\mathbb{C}}$ by stereographic projection means that we define the elliptic metric $d_{\hat{\mathbb{C}}}$ on $\hat{\mathbb{C}}$ by the following.

$$d_{\hat{\mathbb{C}}}(p, q) := d_{S^2}(s^{-1}(p), s^{-1}(q)) \qquad (3.4.7)$$

Proposition 3.4.19 *The elliptic metric (3.4.7) is invariant under the action of the elliptic group. That is, we have*

$$d_{\hat{\mathbb{C}}}(p, q) = d_{\hat{\mathbb{C}}}(Tp, Tq) \qquad (3.4.8)$$

for all $p, q \in \hat{\mathbb{C}}$ and for all $T \in$ ELL.

In order to obtain a formula for computing $d_{\hat{\mathbb{C}}}(p, q)$, we follow the same procedure for hyperbolic distance. First, we find the distance $d_{\hat{\mathbb{C}}}(0, u)$, where $0 \le u \le 1$. Let $S = (0, 0, -1) = s^{-1}(0)$ and let $U = s^{-1}(u)$. Let $0 = (0, 0, 0)$, let $N = (0, 0, 1)$, let $\alpha = \angle SNU$ and let $\theta = \angle SOU$ (see Figure 3.4.20, p. 85). It is a simple exercise to show that $\alpha = \theta/2$, so that we have

$$d_{\hat{\mathbb{C}}}(0, u) = d_{S^2}(S, U) = \theta = 2\alpha = 2\arctan u. \qquad (3.4.9)$$

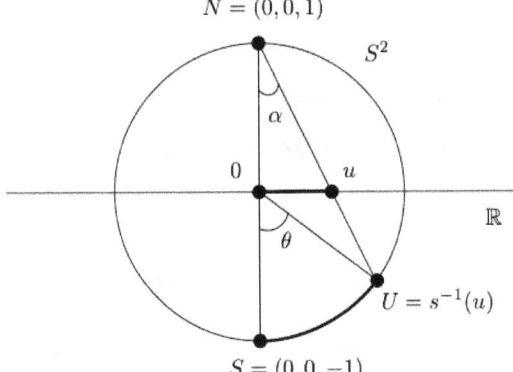

Figure 3.4.20 Transferring the natural metric on S^2 to $\hat{\mathbb{C}}$ via stereographic projection.

Checkpoint 3.4.21 Show that $\theta/2 = \alpha = \arctan u$ in Figure 3.4.20, p. 85.

For the general case, let $z_1, z_2 \in \hat{\mathbb{C}}$, and let $Tz = e^{it} \frac{z - z_1}{z_1^* z + 1}$ be the transformation in ELL (using the form (3.4.5)) that sends $z_1 \to 0$ and $z_2 \to u \geq 0$. Applying (3.4.9), we have the elliptic distance formula in $\hat{\mathbb{C}}$.

$$d_{\hat{\mathbb{C}}}(z_1, z_2) = 2 \arctan \left| \frac{z_2 - z_1}{z_1^* z_2 + 1} \right| \tag{3.4.10}$$

Now let γ be a parametric curve $z(t) = x(t) + iy(t)$ in $\hat{\mathbb{C}}$. Using the same argument as in the paragraph preceding the hyperbolic length integral formula (3.3.6), using the first order Taylor approximation $\arctan u \approx u$ and making the appropriate changes, we arrive at the elliptic length integral formula.

$$\text{Length}(\gamma) = 2 \int_a^b \frac{|z'(t)|}{1 + |z(t)|^2} \, dt \tag{3.4.11}$$

Checkpoint 3.4.22 Show that the first order Taylor approximation for $\arctan u$ is u. Complete the details of modifying the derivation of the hyperbolic length integral formula to obtain the elliptic length integral formula.

Checkpoint 3.4.23 Find the length of the elliptic cycle parameterized by $z(t) = \alpha e^{it}$ for $0 \leq t \leq 2\pi$, where $0 < \alpha \leq 1$.

Using the same argument as in the paragraph preceding the hyperbolic area integral formula (3.3.7), using the elliptic length differential $ds = \frac{2|z'(t)| \, dt}{1 - |z(t)|^2}$ in place of the hyperbolic length differential and making the appropriate changes, we obtain the elliptic area integral formula.

$$\text{Area}(R) = \iint_R dA = \iint_R \frac{4r \, dr \, d\theta}{(1 + r^2)^2} \tag{3.4.12}$$

Checkpoint 3.4.24 Find the area of the elliptic disk $\{|z| \leq \alpha\}$, for $0 < \alpha < 1$.

3.4.7 Exercises

1. Show that the three models of elliptic geometry are equivalent.
2. Prove Proposition 3.4.19, p. 85.

Area of an elliptic triangle. The following sequence of exercises establishes the area formula for elliptic triangles.

3. **Area of an elliptic 2-gon.** An **elliptic 2-gon** is a figure with two vertices connected by two elliptic line segments. In $\hat{\mathbb{C}}$, any 2-gon is congruent to a set of the form $R_\alpha := \{z \in \mathbb{C} : 0 \leq \arg z \leq \alpha\} \cup \{\infty\}$ for some α in the range $0 \leq \alpha < 2\pi$. See Figure 3.4.25, p. 87. Use an area integral in $\hat{\mathbb{C}}$ to show that the area of R_α is 2α. Verify that this is the right answer using a picture of S^2.

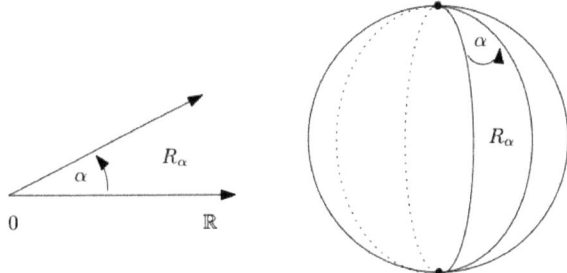

Figure 3.4.25 2-gons in $\hat{\mathbb{C}}$ and S^2.

4.   **Area of an elliptic triangle.** Let $\triangle ABC$ be an elliptic triangle. Let $\mathcal{C}_{AB}, \mathcal{C}_{AC}, \mathcal{C}_{BC}$ denote the great circles that extend the sides AB, AC, BC of the triangle. See Figure 3.4.26, p. 88.

   (a) Explain why A, A' are endpoints of the same diameter, that is, endpoints of a diameter of S^2 or their stereographic projections in $\hat{\mathbb{C}}$.

   (b) Explain why $\triangle A'B'C'$ has the same area as $\triangle ABC$, even though the two triangles are not necessarily congruent! (Note that the interior of $\triangle A'B'C'$ is the *exterior* of the three great circles, that is, on the side that contains the point ∞.) Hint: What does part (a) of this problem imply about the relationship between points X and X' on S^2 for $X = A, B, C$?

   (c) Let R denote the interior of the region shown in the figure on the right in Figure 3.4.26, p. 88. Explain why the area of R is

$$2\angle A + 2\angle B + 2\angle C - 2 \text{ Area}(\triangle ABC).$$

   Suggestion: Decompose R using overlapping 2-gons.

   (d) Let R' denote the exterior of the region R, that is, $R' = \hat{\mathbb{C}} \setminus R$. Explain why the area of R' is

$$2\angle A + 2\angle B + 2\angle C - 2 \text{ Area}(\triangle A'B'C').$$

   Suggestion: Decompose R' into overlapping 2-gons.

   (e) Explain why the area of elliptic triangle $\triangle ABC$ is

$$\text{Area}(\triangle ABC) = (\angle A + \angle B + \angle C) - \pi. \qquad (3.4.13)$$

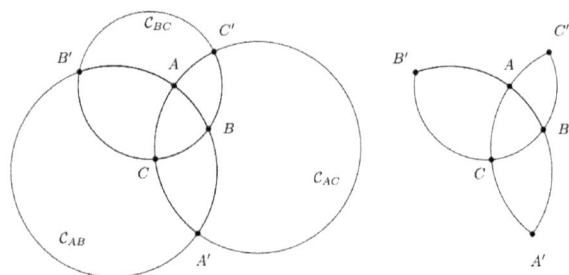

Figure 3.4.26

3.5 Projective geometry

Early motivation for the development of projective geometry came from artists trying to solve practical problems in perspective drawing and painting. In this section, we present a modern Kleinian version of projective geometry.

Throughout this section, $\mathbb{F}$ is a field, V is a vector space over $\mathbb{F}$, $\mathbb{P}(V) = (V \setminus \{0\})/\mathbb{F}^*$ is the projective space, and $PGL(V) = GL(V)/\mathbb{F}^*$ is the projective transformation group. See Exercise 2.5.7, p. 44 for definitions and details. We will write $[T]$ for the projective transformation that is the equivalence class of the linear transformation T of V.

3.5.1 Projective points, lines, and flats

Points in projective space correspond bijectively to 1-dimensional subspaces of V via

$$[v] \leftrightarrow \{\alpha v \colon \alpha \in \mathbb{F}\}.$$

The set of 1-dimensional subspaces in V, denoted $G(1, V)$, is an alternative model space for projective geometry. We will usually denote points in projective space using capital letters, such as P, Q, etc.

A **line** in projective space is a set of the form

$$\ell_\Pi = \{[v] \colon v \in \Pi \setminus \{0\}\}$$

for some 2-dimensional subspace Π in V. Thus, projective lines correspond bijectively to 2-dimensional subspaces of V via

$$\ell_\Pi \leftrightarrow \Pi.$$

The set of 2-dimensional subspaces in V is denoted $G(2, V)$. Points in projective space are called **collinear** if they lie together on a projective line. We will usually denote projective lines using lower case letters, such as ℓ, m, etc.

There is an offset by 1 in the use of the word "dimension" in regards to subsets of $\mathbb{P}(V)$ and the corresponding subspace in V. In general, a **k-dimensional flat** in $\mathbb{P}(V)$ is a set of the form $\{[v] \colon v \in G(k+1, V)\}$, where $G(d, V)$ denotes the set of d-dimensional subspaces of V.[1] Flats are also called *subspaces* in projective space, even though projective space is not a vector space.

Points $P_1 = [v_1], P_2 = [v_2], \ldots, P_k = [v_k]$ are said to be in **general position** if the vectors $v_1, v_2, \ldots, v_k$ are independent in V.

---

[1]The set $G(d, V)$ is called the **Grassmannian** of d-dimensional subspaces of V, named in honor of Hermann Grassmann.

3.5.2 Coordinates

For the remainder of this section, we consider $V = \mathbb{F}^{n+1}$. For readability, we will write $P = [v] = [x_0, x_1, x_2, \ldots, x_n]$ (rather than the more cumbersome $[(x_0, x_1, x_2, \ldots, x_n)]$) to denote the point in projective space that is the projective equivalence class of the point $v = (x_0, x_1, x_2, \ldots, x_n)$ in $\mathbb{F}^{n+1}$. The entries x_i are called **homogeneous coordinates** of P. If $x_0 \neq 0$, then

$$P = [x_0, x_1, x_2, \ldots, x_n] = \left[1, \frac{x_1}{x_0}, \frac{x_2}{x_0}, \ldots, \frac{x_n}{x_0}\right].$$

The numbers x_i/x_0 for $1 \leq i \leq n$ are called **inhomogeneous coordinates** for P. The n degrees of freedom that are apparent in inhomogeneous coordinates explain why $\mathbb{P}(\mathbb{F}^{n+1})$ is called n-dimensional. Many texts write $\mathbb{FP}(n)$, $\mathbb{FP}_n$, or simply $\mathbb{P}_n$ when $\mathbb{F}$ is understood, to denote $\mathbb{P}(\mathbb{F}^{n+1})$.

3.5.3 Freedom in projective transformations

In an n-dimensional vector space, any n independent vectors can be mapped to any other set of n independent vectors by a linear transformation. Therefore it seems a little surprising that in n-dimensional projective space $\mathbb{FP}_n = \mathbb{P}(\mathbb{F}^{n+1})$, it is possible to map any set of $n + 2$ points to any other set of $n + 2$ points, provided both sets of points meet sufficient "independence" conditions. This subsection gives the details of this result, called the Fundamental Theorem of Projective Geometry.

Let $e_1, e_2, \ldots, e_n, e_{n+1}$ denote the standard basis vectors for $\mathbb{F}^{n+1}$ and let $e_0 = \sum_{i=1}^{n+1} e_i$. Let $v_1, v_2, \ldots, v_{n+1}$ be another basis for $\mathbb{F}^{n+1}$ and let $c_1, c_2, \ldots, c_{n+1}$ be nonzero scalars. Let T be the linear transformation T of $\mathbb{F}^{n+1}$ given by $e_i \to c_i v_i$ for $1 \leq i \leq n+1$. Projectively, $[T]$ sends $[e_i] \to [v_i]$ and $[e_0] \to [\sum_i c_i v_i]$.

Now suppose there is another map $[S]$ that agrees with $[T]$ on the $n + 2$ points $[e_0], [e_1], [e_2], \ldots, [e_{n+1}]$. Then $[U] := [S]^{-1} \circ [T]$ fixes all the points $[e_0], [e_1], [e_2], \ldots, [e_{n+1}]$. This means that $Ue_i = k_i e_i$ for some nonzero scalars $k_1, k_2, \ldots, k_{n+1}$ and that $Ue_0 = k' e_0$ for some $k' \neq 0$. This implies

$$(k_1, k_2, \ldots, k_{n+1}) = (k', k', \ldots, k')$$

so we have $k' = k_1 = k + 2 = \cdots k_{n+1}$. Therefore $[U]$ is the identity transformation, so $[S] = [T]$. We have just proved the following existence and uniqueness lemma.

Lemma 3.5.1 *Let $v_1, v_2, \ldots, v_{n+1}$ be an independent set of vectors in $\mathbb{F}^{n+1}$ and let $v_0 = \sum_{i=1}^{n+1} c_i v_i$ for some nonzero scalars $c_1, c_2, \ldots, c_{n+1}$. There exists a unique projective transformation that maps $[e_i] \to [v_i]$ for $0 \leq i \leq n + 1$.*

Theorem 3.5.2 Fundamental Theorem of Projective Geometry.
Let $P_0, P_1, P_2, \ldots, P_{n+1}$ be a set of $n + 2$ points in $\mathbb{P}(\mathbb{F}^{n+1})$ such that all

subsets of size $n + 1$ are in general position. Let $Q_0, Q_1, Q_2, \ldots, Q_{n+1}$ be another such set. There exists a unique projective transformation that maps $P_i \to Q_i$, $0 \le i \le n+1$.

3.5.4 The real projective plane

The remainder of this section is devoted to the planar geometry $\mathbb{P}(\mathbb{R}^3) = \mathbb{RP}_2$ called the **real projective plane**. It is of historical interest because of its early practical use by artists. Lines through the origin in $\mathbb{R}^3$ model sight lines in the real world as seen from an eye placed at the origin. A plane that does not pass through the origin models the "picture plane" of the artist's canvas. Figure 3.5.3, p. 91 shows a woodcut by Albrecht Dürer that illustrates a "perspective machine" gadget used by 16th century artists to put the projective model into practice for image making.

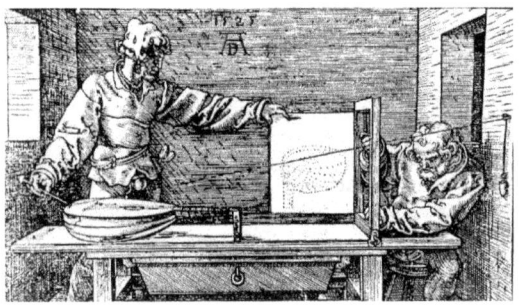

Figure 3.5.3 *Man drawing a lute*, Albrecht Dürer, 1525 (Wikipedia Commons).

A two dimensional subspace Π in $\mathbb{R}^3$ is specified by a normal vector $n = (n_1, n_2, n_3)$ via the equation $n \cdot v = 0$, that is, a point $v = (x, y, z)$ lies on Π with normal vector n if and only if $n \cdot v = n_1 x + n_2 y + n_3 z = 0$. Any nonzero multiple of n is also a normal vector for Π, so the set $G(2, \mathbb{R})$ of 2-dimensional subspaces in $\mathbb{R}^3$ is in one-to-one correspondence with $\mathbb{R}^3/\mathbb{R}^*$. We will write $\ell = [n] = [n_1, n_2, n_3]$ to denote the projective line ℓ whose corresponding 2-dimensional subspace in $\mathbb{R}^3$ has normal vectors proportional to (n_1, n_2, n_3). Beware the overloaded notation! Whether the equivalence class $[v]$ of a vector v in $\mathbb{R}^3$ denotes a projective point or a projective line has to be specified.

The equation $n \cdot v = 0$ makes sense *projectively*. This means that if $n \cdot v = 0$ for vectors n, v, then

$$(\alpha n) \cdot (\beta v) = 0 \quad \text{for all} \quad \alpha, \beta \in \mathbb{F}^*, \tag{3.5.1}$$

even though the value of the dot product is *not* well-defined for projective equivalence classes! Thus we will write $\ell \cdot P = [n_1, n_2, n_3] \cdot [x, y, z] = 0$ for a projective line $\ell = [n_1, n_2, n_3]$ and a projective point $P = [x, y, z]$, to mean (3.5.1), and we make the following interpretation of the dot product as an *incidence relation* in $\mathbb{RP}_2$.

$$\ell \cdot P = 0 \quad \Leftrightarrow \quad P \text{ lies on } \ell \quad \Leftrightarrow \quad \ell \text{ contains } P \qquad (3.5.2)$$

Given two independent vectors v, w in $\mathbb{R}^3$, their cross product $v \times w$ is a normal vector for the 2-dimensional space spanned by v, w. Given two 2-dimensional subspaces Π, Σ in $\mathbb{R}^3$ with normal vectors n, m, the cross product $n \times m$ is a vector that lies along the 1-dimensional subspace $\Pi \cap \Sigma$. The bilinearity of cross product implies that cross product is well-defined on projective classes, i.e., we can write $[u] \times [v] := [u \times v]$. Thus we have the following.

Proposition 3.5.4 *Given two points $P = [u], P' = [u']$ in $\mathbb{RP}_2$, there is a unique projective line $\overline{PP'} = [u \times u']$ that contains them. Given two lines $\ell = [n], \ell' = [n']$ in $\mathbb{RP}_2$, there is a unique projective point $[n \times n']$ in their intersection $\ell \cap \ell'$.*

3.5.5 Exercises

1. Use Lemma 3.5.1, p. 90 to prove the Fundamental Theorem of Projective Geometry (Theorem 3.5.2, p. 90).

2. **Coordinate charts and inhomogeneous coordinates.** To facilitate thinking about the interplay between the projective geometry $\mathbb{P}(\mathbb{F}^{n+1}) = \mathbb{FP}_n$ and the geometry of $\mathbb{F}^n$ (rather than $\mathbb{F}^{n+1}$!) it is useful to have a careful definition for "taking inhomogeneous coordinates in position i". Here it is: Let U_i be the subset of $\mathbb{FP}_n$ of points whose homogeneous coordinate x_i is nonzero. Let $\pi_i \colon U_i \to \mathbb{F}^n$ be given by

$$\pi_i([x_0, x_1, x_2, \ldots, x_n]) = \left(\frac{x_0}{x_i}, \frac{x_1}{x_i}, \frac{x_2}{x_i}, \ldots \frac{x_{i-1}}{x_i}, \frac{x_{i+1}}{x_i}, \ldots, \frac{x_n}{x_i} \right).$$

    The one-sided inverse $\mathbb{F}^n \to \mathbb{FP}_n$ given by

$$(x_0, x_1, \ldots x_{i-1}, \widehat{x_i}, x_{i+1}, \ldots, x_n) \to [x_0, x_1, \ldots x_{i-1}, 1, x_{i+1}, \ldots, x_n]$$

    (where the circumflex hat indicates a deleted item from a sequence) is called the i-th **coordinate chart** for $\mathbb{FP}_n$. What is the map that results from applying the 0-th coordinate chart $\mathbb{C} \to \mathbb{CP}_1 = \mathbb{P}(\mathbb{C}^2)$ followed by taking inhomogeneous coordinates in position 1?

3. **Möbius geometry is projective geometry.** Show that Möbius geometry $(\hat{\mathbb{C}}, \mathrm{MOB})$ and the projective geometry $(\mathbb{P}(\mathbb{C}^2), PGL(2))$ are equivalent via the map $\mu \colon \mathbb{P}(\mathbb{C}^2) \to \hat{\mathbb{C}}$ given by

$$\mu([\alpha, \beta]) = \begin{cases} \alpha/\beta & \beta \neq 0 \\ \infty & \beta = 0. \end{cases} \qquad (3.5.3)$$

Comment: Observe that μ is an extension of $\pi_1 \colon U_1 \to \mathbb{C}$ given by $\pi_1([x_0, x_1]) = \frac{x_0}{x_1}$ (defined in Exercise 3.5.5.2, p. 92).

4. **Cross ratio.** The projective space $\mathbb{P}_1 = \mathbb{P}(\mathbb{F}^2)$ is called the **projective line**. The map $\mu \colon \mathbb{P}_1 \to \hat{\mathbb{F}}$, given by $\mu([x_0, x_1]) = \frac{x_0}{x_1}$ (defined in Exercise 3.5.5.3, p. 92, but where $\mathbb{F}$ may be any field, with $\hat{\mathbb{F}} = \mathbb{F} \cup \{\infty\}$) takes the points

$$[e_0] = [1, 1], [e_2] = [0, 1], [e_1] = [1, 0]$$

in $\mathbb{P}_1$ to the points $1, 0, \infty$ in $\hat{\mathbb{F}}$, respectively. Let $(\cdot, P_1, P_2, P_3)$ denote the unique projective transformation $[T]$ that takes P_1, P_2, P_3 to $[e_0], [e_2], [e_1]$. The **cross ratio** (P_0, P_1, P_2, P_3) is defined to be $\mu([T](P_0))$.

   (a) Show that this definition of cross ratio in projective geometry corresponds to the cross ratio of Möbius geometry for the case $\mathbb{F} = \mathbb{C}$, via the map μ, that is, show that the following holds.

$$(P_0, P_1, P_2, P_3) = (\mu(P_0), \mu(P_1), \mu(P_2), \mu(P_3))$$

   (b) Show that

$$(P_0, P_1, P_2, P_3) = \frac{\det(P_0 P_2) \det(P_1 P_3)}{\det(P_1 P_2) \det(P_0 P_3)}$$

   where $\det(P_i P_j)$ is the determinant of the matrix $\begin{bmatrix} a_i & a_j \\ b_i & b_j \end{bmatrix}$,

   where $P_i = [a_i, b_i]$.

5. **Condition for collinearity in** $\mathbb{RP}_2$. Let $u = (u_1, u_2, u_3)$, $v = (v_1, v_2, v_3)$, $w = (w_1, w_2, w_3)$ be vectors in $\mathbb{R}^3$, and let M be the matrix

$$M = \begin{bmatrix} u_1 & v_1 & w_1 \\ u_2 & v_2 & w_2 \\ u_3 & v_3 & w_3 \end{bmatrix}.$$ Show that $[u], [v], [w]$ are collinear in $\mathbb{RP}_2$ if and only if $\det M = 0$.

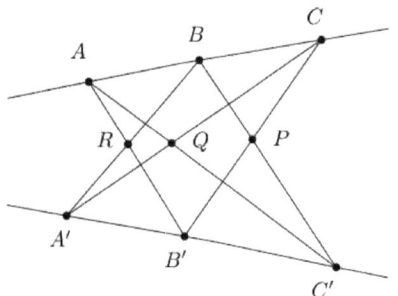

Figure 3.5.5 Pappus' Theorem

6.   The following is a famous theorem of classical geometry.

> **Pappus' Theorem.**
>
> Let A, B, C be three distinct collinear points in $\mathbb{RP}_2$. Let A', B', C' be another three distinct collinear points on a different line. Let P, Q, R be the intersection points $P = BC' \cap B'C$, $Q = AC' \cap A'C$, $R = AB' \cap A'B$. Then points P, Q, R are collinear. See Figure 3.5.5, p. 94.

Follow the outline below to prove Pappus' Theorem under the additional assumption that no three of A, A', P, R are collinear. Applying the Fundamental Theorem of Projective Geometry, we may assume $A = [e_1]$, $A' = [e_2]$, $P = [e_3]$, and $R = [e_0]$.

- Check that $AR = [0, -1, 1]$ and $A'R = [1, 0, -1]$.

- Explain why it follows that $B' = [r, 1, 1]$ and $B = [1, s, 1]$ for some r, s.

- Explain why $C = [rs, s, 1]$ and $C' = [r, rs, 1]$.

- Explain why $Q = [rs, rs, 1]$.

- Observe that P, Q, R all lie on $[1, -1, 0]$.

Hint.   For the second bullet point, use the fact that $B' = [x, y, z]$ lies on AR to get $y = z$. For the third bullet point, use known coordinates for A, B, B', P to get coordinates for lines AB, PB'. Then $C = AB \cap PB'$. Use a similar process for C'. For the fourth bullet point, use $Q = AC' \times A'C$.

7.   **Quadrics.** A **quadric** in $\mathbb{P}(\mathbb{F}^{n+1})$ is a set of points whose homogeneous coordinates satisfy an equation of the form

$$\sum_{0 \le i \le j \le n} c_{ij} x_i x_j = 0. \qquad (3.5.4)$$

A quadric in $\mathbb{RP}_2$ is called a **conic**.

(a) Explain why (3.5.4) is a valid definition of a set of points in $\mathbb{P}(\mathbb{F}^{n+1})$.

(b) Consider the conic C given by

$$x_0^2 + x_1^2 - x_2^2 = 0.$$

What are the figures in $\mathbb{R}^2$ that result from taking inhomogeneous coordinates (see Exercise 3.5.5.2, p. 92) on C in positions $0, 1, 2$?

3.6 Additional exercises

Exercises

1. **Euclidean subgroup of the Möbius group.** Let EUC denote the subgroup of the Möbius group MOB generated by rotations and translations, that is, transformations of the type $z \to e^{it}z$ for $t \in \mathbb{R}$ and $z \to z + b$ for $b \in C$. The geometry $(\mathbb{C}, \text{EUC})$ is sometimes called "Euclidean geometry". Is $(\mathbb{C}, \text{EUC})$ equivalent to the Euclidean geometry defined in Subsection 3.1.1, p. 50? Why or why not?

2. **Elliptic geometry and spherical geometry.** Is elliptic geometry $(\hat{\mathbb{C}}, \text{ELL})$ equivalent to spherical geometry defined in Subsection 3.1.1, p. 50? Why or why not?

3. **Parallel displacements in hyperbolic geometry.** Let T be an element of the hyperbolic group HYP, with a single fixed point p and normal form

$$\frac{1}{Tz - p} = \frac{1}{1 - p} + \beta.$$

   Show that $p\beta$ must be a pure imaginary number, that is, there must be a real number k such that $p\beta = ki$.

4. Prove that all elements of the elliptic group ELL are elliptic in the normal form sense, i.e., we have $|\alpha| = 1$ in the normal form expression

$$\frac{Tz - p}{Tz - q} = \alpha \frac{z - p}{z - q}.$$

   Suggestion: First find the fixed points p, q, then put $z = \infty$ in the normal form equation and solve for α.

5. **Alternative derivation of the formula for elliptic group elements.** To obtain an explicit formula for elements of the elliptic group, we begin with a necessary condition. Let $R = s^{-1} \circ T \circ s$ be the rotation of S^2 that lifts T via stereographic projection. If P, Q are a pair of endpoints of a diameter of S^2, then $R(P), R(Q)$ must also be a pair of endpoints of a diameter. Exercise 1.3.4.7, p. 12 establishes the condition that two complex numbers p, q are stereographic projections of endpoints of a diameter if and only if $pq^* = -1$. Thus we have the following necessary condition for T.

$$pq^* = -1 \quad \text{implies} \quad Tp(Tq)^* = -1 \tag{3.6.1}$$

   Now suppose that $Tz = \frac{az+b}{cz+d}$ with $ad - bc = 1$. Solving the equation $Tp = \frac{-1}{(T(\frac{-1}{p^*}))^*}$ leads to $c = -b^*$ and $d = a^*$. Thus we conclude that T has the following form.

$$Tz = \frac{az + b}{-b^*z + a^*}, \quad |a|^2 + |b|^2 = 1 \tag{3.6.2}$$

Carry out the computation to derive (3.6.2). Explain why there is no loss of generality by assuming $ad - bc = 1$.

6. **Identifications of $U(\mathbb{H})$ and ELL with $\text{Rot}(S^2)$.** The discussion of elliptic geometry (Section 3.4, p. 78) establishes two ways to construct rotations from matrices. The purpose of this exercise is to reconcile these identifications. Given $a, b \in \mathbb{C}$ with $|a|^2 + |b|^2 = 1$, let us define the following objects, all parameterized by a, b.

$$M_{a,b} = \begin{bmatrix} a & b \\ -b^* & a^* \end{bmatrix}$$

$$r_{a,b} = \text{Re}(a) + \text{Im}(a)i + \text{Re}(b)j + \text{Im}(b)k$$

$$R_{r_{a,b}} = \left[u \to r_{a,b} u r_{a,b}^*\right] \quad \text{for } u \in S_{\mathbb{H}}^2$$

$$T_{a,b} = \left[z \to \frac{az + b}{-b^* z + a^*}\right] \quad \text{for } z \in \hat{\mathbb{C}}$$

$$R_{T_{a,b}} = s^{-1} \circ T_{a,b} \circ s$$

The above objects are organized along two sequences of mappings. The rotation $R_{r_{a,b}}$ is at the end of the "quaternion path"

$$SU(2) \to U(\mathbb{H}) \to \text{Rot}(S_{\mathbb{H}}^2) \qquad (3.6.3)$$
$$M_{a,b} \to r_{a,b} \to R_{r_{a,b}}$$

and the rotation $R_{T_{a,b}}$ is at the end of the "Möbius path"

$$SU(2) \to \text{ELL} \to \text{Rot}(S^2) \qquad (3.6.4)$$
$$M_{a,b} \to T_{a,b} \to R_{T_{a,b}}.$$

This problem is about comparing the rotations $R_{r_{a,b}}$ and $R_{T_{a,b}}$ (see Table 3.6.1, p. 97) and reconciling the difference. The angles of rotation are the same, but the axes are different, but only by a reordering of coordinates and a minus sign.

Table 3.6.1 Axis and angle of rotation for quaternion and Möbius rotation constructions

$R_{v,\theta}$	axis of rotation v	angle of rotation θ
$R_{r_{a,b}}$	$\dfrac{(\text{Im}(a),\text{Re}(b),\text{Im}(b))}{\sqrt{1-(\text{Re}(a))^2}}$	$2\arccos(\text{Re}(a))$
$R_{T_{a,b}}$	$\dfrac{(\text{Im}(b),-\text{Re}(b),\text{Im}(a))}{\sqrt{1-(\text{Re}(a))^2}}$	$2\arccos(\text{Re}(a))$

The exercises outlined below verify the values for v, θ in Table 3.6.1, p. 97.

(a) Use Proposition 1.2.9, p. 8 to justify the values for v and θ for $R_{r_{a,b}}$ in Table 3.6.1, p. 97.

(b) Solve $T_{a,b}z = z$ to show that one of the fixed points of $T_{a,b}$ is

$$p = -ib\left(\frac{\sqrt{1-(\mathrm{Re}(a))^2} + \mathrm{Im}(a)}{|b|^2}\right).$$

(c) Show that $s\left(\frac{(\mathrm{Im}(b), -\mathrm{Re}(b), \mathrm{Im}(a))}{\sqrt{1-(\mathrm{Re}(a))^2}}\right) = p$.

(d) Show that

$$T_{a,b} = s \circ h \circ R_{r_{a,b}} \circ h \circ s^{-1} \qquad (3.6.5)$$

where $h \colon \mathbb{R}^3 \to \mathbb{R}^3$ is given by $(x, y, z) \to (z, -y, z)$. Here's one way to do this: evaluate both sides of (3.6.5) on the three points $p, 0, \infty$. Explain why this is sufficient! Use quaternion multiplication to evaluate $R_{r_{a,b}}$. For example, $R_{r_{a,b}}(1, 0, 0) = r_{a,b}ir_{a,b}^*$ under the natural identification $\mathbb{R}^3 \leftrightarrow \mathbb{R}_{\mathbb{H}}^3$.

(e) Here is one way to reconcile the quaternion path (3.6.3) with the Möbius path (3.6.4). Let $H = \frac{1}{\sqrt{2}}\begin{bmatrix} 1 & 1 \\ 1 & -1 \end{bmatrix}$ (the matrix H is sometimes called the *Hadamard* matrix) and let C_{iH} denote the map $M \to (iH)M(iH)^{-1}$. Show that the diagram in Figure 3.6.2, p. 98 commutes. Hint: Notice that $iH \in SU(2)$ and that $Q(iH) = \frac{1}{\sqrt{2}}(i + k)$, and that $R_{Q(iH)} = h$.

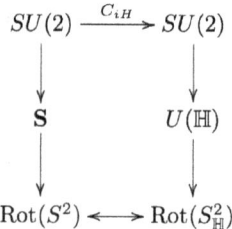

Figure 3.6.2 The map C_{iH} is given by $C_{iH}(T) = (iH)T(iH)^{-1}$. The column of maps on the left is the "Möbius path", and the column of maps on the right is the "quaternion path".

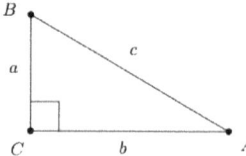

Figure 3.6.3 Right triangle $\triangle ABC$

7. **Pythagorean Theorems.** Let $\triangle ABC$ be a right triangle with right angle $\angle C$ with side lengths $a = d(B, C)$, $b = d(A, C)$, and $c = d(A, B)$ so that the length of the hypotenuse is c. See Figure 3.6.3, p. 98.

(a) Prove the following identities.

$$\cosh\left(\ln\left(\frac{1 + u}{1 - u}\right)\right) = \frac{1 + u^2}{1 - u^2} \quad (0 < u < 1) \qquad (3.6.6)$$

$$\cos(2 \arctan u) = \frac{1 - u^2}{1 + u^2} \qquad (3.6.7)$$

(b) **The Hyperbolic Pythagorean Theorem.** Show that

$$\cosh(c) = \cosh(a)\cosh(b) \qquad (3.6.8)$$

if T is a hyperbolic triangle.

(c) **The Elliptic Pythagorean Theorem.** Show that

$$\cos(c) = \cos(a)\cos(b) \qquad (3.6.9)$$

if T is an elliptic triangle.

Suggestion: Use a transformation to place C at 0 in $\mathbb{D}$ or $\hat{\mathbb{C}}$, with A real and B pure imaginary. Use the formula $d(p, q) = \ln((1 + u)/(1 - u))$ with $u = \left|\frac{q - p}{1 - p^* q}\right|$ for hyperbolic distance. Use the formula $d(p, q) = 2 \arctan(u)$, with $u = \left|\frac{q - p}{1 + p^* q}\right|$ for elliptic distance. The identities from part (a) will be useful.

Appendix A

Further topics

Here are some topics and sources for independent study projects that are accessible at the level of this textbook.

- The classification of the Platonic solids using group theory [8]
- More group theory with connections to geometry [2]
- More group theory and an introduction to rings and fields [3]
- More geometry using a Kleinian approach [4]
- An elementary introduction to the Hopf fibration [6]
- Algebra and geometry in quantum mechanics [7]

Appendix B

Solutions to Some Exercises

1 · Preliminaries
1.3 · Stereographic projection
1.3.4 · Exercises

Formulas for inverse stereographic projection.
   1.3.4.1. Answer.

$$s^{-1}(r) = \begin{cases} \left(\frac{2r}{r^2+1}, \frac{r^2-1}{r^2+1} \right) & \text{if } r \neq \infty \\ (0,1) & \text{if } r = \infty \end{cases}$$

$$s^{-1}(3) = (3/5, 4/5)$$

   1.3.4.2. Answer.

$$s^{-1}(z) = \begin{cases} \left(\frac{2\operatorname{Re}(z)}{|z|^2+1}, \frac{2\operatorname{Im}(z)}{|z|^2+1}, \frac{|z|^2-1}{|z|^2+1} \right) & \text{if } z \neq \infty \\ (0,0,1) & \text{if } z = \infty \end{cases}$$

$$s^{-1}(3+i) = (6/11, 2/11, 9/11)$$

2 · Groups
2.1 · Examples of groups
2.1.1 · Permutations

Checkpoint 2.1.2 Answer. $\sigma\tau = [1,3,2]$, $\tau\sigma = [3,2,1]$, $\sigma^2 = [2,3,1]$

2.1.2 · Symmetries of regular polygons

Checkpoint 2.1.6 Answer.

   1. $R_{3/4}$

2. $F_{D'}$

3. F_D

4. $R_{1/4}$

5. $R_{3/4}$

6. R_0

7. $F_{D'}$

2.2 · Definition of a group

· **Exercises**

2.2.10. Cayley tables.

Answer 1.

	e	(23)	(13)	(12)	(123)	(132)
e	e	(23)	(13)	(12)	(123)	(132)
(23)	(23)	e	(123)	(132)	(13)	(12)
(13)	(13)	(132)	e	(123)	(12)	(23)
(12)	(12)	(123)	(132)	e	(23)	(13)
(123)	(123)	(12)	(23)	(13)	(132)	e
(132)	(132)	(13)	(12)	(23)	e	(123)

Answer 2.

	F_V	F_H	F_D	$F_{D'}$	$R_{1/4}$	$R_{1/2}$	$R_{3/4}$	R_0
F_V	R_0	$R_{1/2}$	$R_{3/4}$	$R_{1/4}$	$F_{D'}$	F_H	F_D	F_V
F_H	$R_{1/2}$	R_0	$R_{1/4}$	$R_{3/4}$	F_D	F_V	$F_{D'}$	F_H
F_D	$R_{1/4}$	$R_{3/4}$	R_0	$R_{1/2}$	F_V	$F_{D'}$	F_H	F_D
$F_{D'}$	$R_{3/4}$	$R_{1/4}$	$R_{1/2}$	R_0	F_H	F_D	F_V	$F_{D'}$
$R_{1/4}$	F_D	$F_{D'}$	F_H	F_V	$R_{1/2}$	$R_{3/4}$	R_0	$R_{1/4}$
$R_{1/2}$	F_H	F_V	$F_{D'}$	F_D	$R_{3/4}$	R_0	$R_{1/4}$	$R_{1/2}$
$R_{3/4}$	$F_{D'}$	F_D	F_V	F_H	R_0	$R_{1/4}$	$R_{1/2}$	$R_{3/4}$
R_0	F_V	F_H	F_D	$F_{D'}$	$R_{1/4}$	$R_{1/2}$	$R_{3/4}$	R_0

2.3 · Subgroups and cosets

Checkpoint 2.3.2 Answer.

1. Yes. The composition of any two rotations is a rotation, and every rotation has an inverse that is also a rotation.

2. No. Just observe that $F_H^2 = R_0$ is not a reflection. The group operation on D_4 does not restrict properly to the subset of reflections.

Checkpoint 2.3.3 Answer.

$$G/H = \{eH, (12)H, (13)H, (23)H, (123)H, (132)H\}$$
$$= \{\{e, (12)\}, \{(12), e\}, \{(13), (123)\},$$
$$\{(23), (132)\}, \{(123), (13)\}, \{(132), (23)\}\}$$
$$= \{H, \{(13), (123)\}, \{(23), (132)\}\}$$

Checkpoint 2.3.8 Answer.

1. $\langle F_H, F_V \rangle = \{R_0, R_{1/2}, F_H, F_V\}$

2. $\langle 6, 8 \rangle = \langle 2 \rangle = 2\mathbb{Z}$

· **Exercises**

2.3.2. Answer. In the "list of values" permutation notation of Checkpoint 2.1.2, p. 22, the subgroups of S_3 are $\{[1,2,3]\}$, $\{[1,2,3], [2,1,3]\}$, $\{[1,2,3], [1,3,2]\}$, $\{[1,2,3], [3,2,1]\}$, $\{[1,2,3], [2,3,1], [3,1,2]\}$, and S_3. In cycle notation, the subgroups of S_3 (in the same order) are $\{e\}$, $\{e, (12)\}$, $\{e, (23)\}$, $\{e, (13)\}$, $\{e, (123), (132)\}$, S_3.

2.4 · Group homomorphisms

Checkpoint 2.4.4 Answer.

1. C_n

2. $\langle 4 \rangle = \{0, 4\}$

3. $\{e\}$

2.5 · Group actions

Checkpoint 2.5.2 Answer.

1. $\text{Orb}((1,1)) = \{(1,1), (1,-1), (-1,1), (-1,-1)\}$, $\text{Orb}((1,0)) = \{(1,0), (-1,0), (0,1), (0,-1)\}$, $\text{Stab}((1,1)) = \{R_0, F_{D'}\}$, $\text{Stab}((1,0)) = \{R_0, F_H\}$

2. $\text{Orb}(\pi/4, \pi/6)$ is the horizontal circle on S^2 with "latitude" $\pi/4$, $\text{Orb}(0,0) = \{(0,0)\}$, $\text{Stab}(\pi/4, \pi/6) = \{1\}$, $\text{Stab}(0,0) = S^1$

3. $\text{Orb}(x) = \{gxg^{-1} : g \in G\}$, the stabilizer of x is the centralizer subgroup $C(x)$ (see Exercise 2.3.5, p. 34)

3 · Geometries
3.2 · Möbius geometry
3.2.2 · The Fundamental Theorem of Möbius Geometry

Checkpoint 3.2.11 Answer.

$$(z_1 = \infty) \qquad\qquad Tz = \frac{z - z_2}{z - z_3}$$

$$(z_2 = \infty) \qquad\qquad Tz = \frac{z_1 - z_3}{z - z_3}$$

$$(z_3 = \infty) \qquad\qquad Tz = \frac{z - z_2}{z_1 - z_2}$$

3.3 · Hyperbolic geometry
3.3.4 · Hyperbolic length and area

Checkpoint 3.3.15 Solution. Let $Sz = \frac{z - z_1}{1 - z_1^* z}$, let $t = -\arg(Sz_2)$, and let $Tz = e^{it} Sz$, so that we have $Tz_1 = 0$ and $Tz_2 = u > 0$. Because $T \in \mathrm{HYP}$, T is determined by the two parameters z_1, t.

References

[1] Lars V. Ahlfors. *Complex Analysis: An Introduction to the Theory of Analytic Functions of One Complex Variable*. McGraw-Hill, 1953.

[2] M.A. Armstrong. *Groups and Symmetry*. Springer, 1988.

[3] Joseph A. Gallian. *Contemporary abstract algebra*. Cengage Learning, 9th edition, 2017.

[4] Michael Henle. *Modern Geometries: Non-Euclidean, Projective, and Discrete*. Prentice Hall, 2 edition, 2001.

[5] David W. Lyons. Complex Numbers. *Not Just Calculus*. mathvista.org (www.mathvista.org), 2024.

[6] David W. Lyons. An elementary introduction to the Hopf fibration. *Mathematics Magazine*, 76(2):87--98, 2003.

[7] Michael A. Nielsen and Isaac L. Chuang. *Quantum Computation and Quantum Information*. Cambridge University Press, 2000.

[8] Hermann Weyl. *Symmetry*. Princeton University Press, Princeton, New Jersey, 1952.

Index

Notation

(Continued on next page)

Symbol	Description	Page
$H \trianglelefteq G$	H is a normal subgroup of G	37
$\operatorname{Aut}(G)$	the group of automorphisms of a group G	39
$\operatorname{Inn}(G)$	group of inner automorphisms of a group G	39
$\operatorname{Orb}(x)$	orbit of x under a group action	40
$\operatorname{Stab}(x)$	stabilizer of an element x under a group action	40
X/G	set of orbits of the action of group G on set X	41
$\operatorname{sgn}(\alpha)$	the sign of a permutation α	41
A_n	the alternating group	42
$\mathbb{P}(V)$	projective space	44
$PGL(V)$	the projective linear group	44
U_n	group of units in $\mathbb{Z}_n$	45
$F \cong F'$	figure F is congruent to figure F'	50
MOB	Möbius transformation group	55
$\mathbb{D}$	the open unit disk	67
HYP	the hyperbolic group	67
$\mathbb{U}$	the upper half-plane	73
$\mathrm{HYP}_\mathbb{U}$	the upper half-plane group	73
$SU(2)$	the special unitary group	78
$R_{v,\theta}$	rotation about the axis v by angle θ	79
ELL	the elliptic group	81

www.ingramcontent.com/pod-product-compliance
Lightning Source LLC
Chambersburg PA
CBHW072141090326
40927CB00004B/246

.